Chemical Applications
of Pattern Recognition

CHEMICAL APPLICATIONS
OF PATTERN RECOGNITION

PETER C. JURS
Associate Professor of Chemistry
The Pennsylvania State University

THOMAS L. ISENHOUR
Professor of Chemistry
University of North Carolina

A WILEY-INTERSCIENCE PUBLICATION

John Wiley & Sons
New York / London / Sydney / Toronto

CHEMISTRY

Library of Congress Cataloging in Publication Data:

Jurs, Peter C.
    Chemical applications of pattern recognition.

    "A Wiley-Interscience publication."
    Bibliography: p.
    Includes index.
    1. Electronic data processing—Chemistry.
2. Pattern perception. I. Isenhour, Thomas L., joint author. II. Title.

QD39.3.E46J87   540'.28'54   74-26720
ISBN 0-471-45330-7

Printed in the United States of America

10 9 8 7 6 5 4 3 2 1

Preface

This book is an introduction to the application of some pattern recognition methods to problems from chemistry. While the field of pattern recognition is extraordinarily broad and encompasses a wide variety of methods, approaches, and applications areas, we have chosen to narrow our focus. We have devoted the largest fraction of this book to the type of nonparametric pattern recognition systems known as "learning machines." They are so named because they increase their ability to produce correct responses for classification tasks as their experience increases. The basic building block of such systems is the adaptive binary pattern classifier or the adaptive threshold logic unit. These units investigate a set of labeled data for invariance which can be linked to the classifications. This procedure is known as supervised training. Most of the work reported in this book employed adaptive binary pattern classifiers which developed their classification capability by supervised training.

The majority of the applications discussed come from spectroscopic data analysis, for example, low-resolution mass spectra, infrared spectra, NMR spectra. In these areas pattern recognition was first used in chemistry. However, more recently new areas within chemistry have been investigated, and it may be that the ability of pattern recognition methods to study a very large class of widely varying chemical problems will in the future be the most useful feature of the technique.

The book is organized into seven chapters. The first chapter deals with the overall pattern recognition system and briefly introduces each component. Chapter II introduces the fundamentals of the notation used for binary pattern classifiers, describes the basic training method employed for their development, and discusses the properties by which the usefulness of binary pattern classifiers can be measured. Chapter III describes several ways of preprocessing data to make classification more economical, and

v

gives examples drawn from studies of spectroscopic and electrochemical data. Chapter IV discusses a variety of discriminant functions and their development in detail. Numerous examples are included from chemical investigations. In Chapter V feature selection driven by results obtained from the classifier is described. Examples from mass spectral and infrared spectral studies are given. Chapter VI presents several complex transformations which can be applied to data prior to classification, for example, the development of terms taking into account interactions between primitive descriptors, Fourier transforms, and factor analysis. Chapter VII introduces the idea that the pattern recognition approach can be used for other classes of problems in chemistry. Here the methods of pattern recognition are used to generate mass spectra of compounds only from their two-dimensional molecular structure. The same techniques for structural descriptions are finding application for numerous diverse problems throughout chemistry, for example, structure-activity studies of drugs. A complete basic learning machine program with a sample set of five-dimensional data and sample output is presented in the Appendix. The program should need few if any changes to be executed on computer systems other than the Penn State University Computation Center IBM 370/168 on which the sample run was made. Finally, a bibliography of pattern recognition books is given. These works vary widely in focus but will provide the interested person with an entrance into the voluminous pattern recognition literature.

We are indebted to the many students and co-workers who have helped in the development of our thinking and have, in many cases, carried out the actual experiments described.

PETER C. JURS
THOMAS L. ISENHOUR

State College, Pennsylvania
Chapel Hill, North Carolina
February 1975

Contents

CHAPTER I

Introduction

The advent of high-speed, stored-program digital computers as powerful generalized information-handling devices has led to revolutionary developments in many fields. Among these are several fields which deal with problems that previously appeared to have only extremely difficult, complex, or even unrealizable solutions. Numerous such problems have proved amenable to attack by pattern recognition techniques. Pattern recognition has become a fertile area for the development of concepts and techniques now being applied routinely to problems formerly considered to be approachable only by humans. Thus pattern recognition has come to be a subset of the artificial intelligence or machine intelligence field.

Pattern recognition includes the detection, perception, and recognition of regularities (invariant properties) among sets of measurements describing objects or events. The purpose of pattern recognition is generally to categorize a sample of observed data as a member of a class to which it belongs. This general approach has been applied to problems from many diverse fields.

PATTERN RECOGNITION APPLICATIONS

An astounding variety of practical problems has been attacked with pattern recognition techniques. Several excellent reviews of the pattern recognition literature have appeared which dramatize the enormous breadth of pattern recognition applications (1–5). In addition, a large number of books has been published as well, as shown in the Bibliography.

Some of the motivations underlying the research effort accorded the pattern recognition area are:

1. Substitution of machines for humans in performing routine information processing tasks. Machines capable of processing information more quickly, accurately, safely, or inexpensively are attractive.

2. The development of maximally effective man-machine interfacing requires the machine to use pattern recognition techniques. The machine must be capable of handling optical patterns and natural language, since these are man's preferred means of communication.

3. Pattern recognition is also an attractive field for research for its own sake. The outstanding research problem here is: in general, what is required (in terms of machine arrangement or organization) to obtain performance in perceptual responses similar to that of humans?

It should be emphasized that research projects in pattern recognition often yield results in all three areas.

The types of data that have been analyzed by pattern recognition methods naturally fall into two categories—well-defined coded structures and pictorial representations. Well-defined coded structures are exemplified by printed letters, script, fingerprints, and other data that are (ideally) binary quantized. Pictorial representations include photographs and video data which must be represented with levels of gray. Data that do not fall easily into either of these two categories have also been studied. The following paragraphs discuss some specific problems that have been studied with pattern recognition.

The term "character recognition" refers to research dealing with the interpretation of printed characters and script (6). The development of automated equipment to read special type fonts, such as the magnetic numerals now widely used on bank checks is the most stylized and most successful approach to this problem. A more difficult problem which has also been studied extensively is the recognition of impact-printed characters, such as typewriter or line printer output. Several data sets of impact-printed material have become standard and have therefore been studied by workers using a variety of pattern recognition methods. The even more difficult problem of recognizing hand-printed alphanumeric characters has been studied as well. An additional motivation behind the search for effective man-machine interfacing by automatic recognition of hand-printed material is the drive toward automation of postal services.

Another area of pattern recognition related to man-machine communication is that of speech recognition. The two subproblems in this area deal with the recognition of what is said or the recognition of the identity of the speaker. The far-reaching utility of machines with capabilities for making such decisions is obvious.

Photographic processing is an area in which problems amenable to pattern recognition approaches abound. Accordingly, the processing of photographic images has received much research attention. Microphotographs of biological materials including blood cells and chromosomes have been interpreted using pattern recognition techniques (7). Aerial photographs have been studied for military purposes and remote-sensing reasons. For example, identification of crop types, and presence of forest fires, drought conditions, or other terrain features can be made from aerial photographs (8). Photographs of events in bubble, spark, or cloud chambers have been interpreted by pattern recognition techniques (9). Fingerprint analysis has been made using pattern recognition (10).

Geological data including aerial photographs and seismic signals have been interpreted using pattern recognition. Medical diagnostics including the interpretation of electrocardiograms (EKGs), electroencephalograms (EEGs), and blood analysis have been developed using pattern recognition methods (11, 12). Weather forecasting based on wind, atmospheric pressure, humidity, and temperature measurements has been studied.

BASIC PATTERN RECOGNITION SYSTEM

A general pattern recognition system must be capable of observing a sample of data, preprocessing and transforming it, and classifying the pattern correctly. The basic configuration of a pattern recognition system is shown in Figure 1. It consists of three interrelated subunits: a transducer, a preprocessor (or feature extractor), and a classifier. Although these three subunits are highly interdependent in any implementation of a pattern recognition system, it is convenient to separate them for pedagogical purposes.

The transducer translates information from the real world into the pattern space of the pattern recognition system. Since pattern recognition systems are normally implemented with computer software, the transducer essentially renders raw data into computer-compatible form. Normally, the computer-compatible form used is a string of scalar measurements comprising an n-tuple called the pattern vector: $\mathbf{X} = x_1, x_2, \ldots, x_n$. Each component of the pattern vector represents a physically measurable quantity.

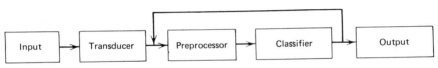

Figure 1. *Basic pattern recognition system.*

The preprocessor's purpose is to accept the pattern vector to be classified and to transform or preprocess it to make the classification task easier. The transformed data are then input to the classifier. The classifier operates on the transformed data to produce a classification decision. More detailed descriptions of the methods used to attain these goals are given in the following discussion.

Transducer

The transducer accepts the raw data of the objects to be classified. It implements a procedure to transform each pattern of raw data into an n-dimensional vector in an n-dimensional euclidean space called the pattern space. The string of measurements forming the pattern vector must contain the essence of the raw input data. Thus the actual implementation of the transducer is entirely dependent on the nature of the raw data. When the raw data consist of time series, such as interferograms or EEGs, a sampling procedure in time might be in order. When they are a function of frequency, for example, infrared spectra, a sampling procedure in frequency is appropriate. When the raw data are pictures, the field can be examined for darker and lighter areas, edges, or geometric forms. Several methods used in reducing pictorial data to a string of measurements can be found in a review article by Levine (5). If the raw data were naturally digital data, such as a low-resolution mass spectrum, the transducer step might be unnecessary. The pattern vectors, or n-dimensional points, developed by the transducer would then be passed to the feature extraction or preprocessing unit of the sytem.

Preprocessor—Feature Extractor

The preprocessor–feature extractor subunit accepts pattern vectors produced by the transducer, and it operates on them in order tp pursue the following goals.

1. To eliminate or at least reduce the fraction of information contained in the raw data that is irrelevant or even confusing. For example, pictorial data should be independent of translations, scale changes, or rotations, and wave forms should be independent of shifts in time or phase.

2. To preserve sufficient information to allow discrimination among the pattern classes, that is, to discover invariances among patterns of common class.

3. To preserve the information in the pattern vector in a form such that it can be effectively utilized by a linear classifier, if possible.

Some specific feature extraction or preprocessing techniques that have been used are mentioned in the following discussion.

The simplest preprocessing involves normalizations and other methods for scaling the components of the pattern vector. One such normalization involves setting the sum of the components of each pattern vector (or their squares) equal to an arbitrary, convenient constant. Another procedure involves allowing the weight of a particular descriptor (dimension) to be inversely proportional to the variance of that descriptor in the pattern space. A more sophisticated procedure involves utilizing a covariance matrix to set up a matrix equation which is solved for eigenvectors and eigenvalues. This yields a set of orthogonal dimensions which define a new pattern space, which is a rotation of the old one, in which classification might be more easily done. This procedure, known as principal components analysis or Karhunen-Loève analysis, can also be used to decrease the dimensionality of pattern vectors. Only the new, transformed dimensions which have large eigenvalues associated with them are saved, while the remainder are discarded. These are just a few of the linear transformation that have been utilized in pattern recognition systems.

Several more complex transformations have also been employed. For example, pattern vectors can be subjected to a Fourier transformation, and then the power spectrum can be developed. Iteratively defined transformations have also been developed. This involves minimizing an error criterion iteratively, for example, the difference between the distances between all pairs of points in the original pattern space and in the new, lower-dimenstional pattern space. Pattern vectors can also be represented by polynomial expansions.

Templates or prototype patterns have been used for comparison with the patterns to be classified to identify important features. Interactive techniques, sometimes involving computer graphics and mapping routines, have been used. Calculations of statistical parameters such as moments and histograms directly from patterns have been made. As is evident from these few paragraphs, the number of feature extraction–preprocessing algorithms is enormous. And it is growing rapidly, since the methods chosen for application to any problem are highly dependent on the problem. Depending on how well it is done, the preprocessing step can lead to success or failure of a pattern recognition study. It is a widely held opinion that the feature extraction–preprocessing stage is the area in which the major new advances will come in the study of pattern recognition systems.

Classifier

The transformed patterns are classified by the third subunit of the pattern recognition system. Classifiers have been developed by utilizing various branches of applied mathematics: statistical decision theory, information theory, geometric theory, and so on.

The task of the classifier can be stated in general as follows. A set of transformed pattern vectors, termed the training set, is used to determine a decision function $f(X)$ such that

$$f(X) = \; >0 \text{ for } X \text{ members of class 1}$$

$$\leq 0 \text{ for } X \text{ members of class 2}$$

The procedure used to develop $f(X)$ is commonly known as the adaptation, training, or learning phase. The goal is to minimize the probability of error in the classifications.

Implementations of $f(X)$ fall naturally into two categories: parametric methods and nonparametric methods. Parametric training methods begin by estimating the statistical parameters of the samples forming the training set. The estimates are subsequently used for specification of the discriminant functions. The most common parametric discriminant function is Bayes' rule, since it is optimum for a well-defined class of problems.

In order to apply Bayes' rule to a two-class pattern recognition problem, the functional form for the conditional density functions for each class and the parameters of these conditional density functions must be known. The usual assumption is that the patterns are normally distributed about their class means. Then the requirement is that the mean vectors and covariance matrices of the classes be known. Additionally, the loss functions that define the severity of misclassifying a pattern must be specified. Then the discriminant function is:

$$f(X) = \frac{P_1 L_1 F_1(X)}{P_2 L_2 F_2(X)} - 1$$

where P_1 = the a priori probability of occurrence of a pattern in class 1; $P_2 = 1 - P_1$ = the a priori probability of occurrence of a pattern in class 2; L_1 and L_2 = the losses associated with misclassifying a member of class 1 or class 2; and $F_1(X)$ and $F_2(X)$ = the probability density functions of class 1 and class 2.

On the assumption that the patterns are gaussian and that the mean vectors and covariance matrices are truly representative of their classes, then the Bayes discriminant function is optimum. Therefore Bayes' criterion has been used as a benchmark for comparison of other discriminant functions.

If the patterns of the training set cannot be described by statistical measures, then a nonparametric discriminant function must be employed. During the development of nonparametric discriminant functions, the only data used are the training set patterns themselves. In order for the training method to yield reliable results, the training set size must be large enough to be representative of the data set from which it is drawn. (A large data set is a necessary, but not sufficient, condition to allow the possibility of development of parametric discriminant functions by the estimation of probability functions.)

A widely studied nonparametric binary pattern classifier is the threshold logic unit (TLU). Besides the pattern vectors of the training set, the only adjustable parameters of TLUs are their linear coefficients which are determined during training. An adaptive TLU is provided with the means to monitor its own performance in relation to a specified index of performance, and it is capable of modifying its own parameters so as to improve its performance. TLUs are usually adaptive only during the design stage. Chapter II deals at length with the characteristics of TLUs. Individual TLUs can be combined in interdependent grids to implement more sophisticated classifiers known as piecewise linear or layered classifiers.

Another popular nonparametric discriminant method is called the K-nearest-neighbor classification. An unknown pattern is classified as a member of the class most often represented among its K nearest neighbors. Nearest is usually defined by a euclidean distance measure, but any metric can be employed. The K-nearest-neighbor classification procedure suffers from the fact that all pattern vectors must be stored and that many computations must be performed to make a classification. However, it has been shown that the error rate of the first nearest-neighbor method is at most twice that of the optimal Bayes' error in which all the underlying probabilities are known (13). Its conceptual and computational simplicity are also appealing. Later investigations with nearest-neighbor classification have shown that a carefully chosen subset of the data sample can be used, thereby reducing storage requirements and computational burden (e.g., 14).

If the data to be classified cannot be described statistically and the classes of the individual pattern vectors are not known, then a different type of classifier is used. These are generally known as clustering algorithms, because it is necessary to allow the pattern vectors themselves to define the number of clusters of points that might then be related to classes. The procedures used have also been called unsupervised training, since the classes of the training samples are unknown. One approach to the difficult task of finding a viable approach to these problems has been to use histograms of the data to approximate the probability density functions of the classes and proceed with other discriminant function methods.

Another nonparametric discriminant method is that of potential functions. Each of the known points in a pattern space is surrounded by its own potential field. To classify an unknown pattern, the overall potential field for each possible class is evaluated, and the class exerting the strongest field is said to be the class of the unknown. This method is extremely powerful, because the function form of the potential as well as the controlling parameters can be varied as a function of class membership, region of pattern space, probability of the presence of different classes, and so on.

These are just a few of the large number of classifiers that have been developed. Additional background information regarding classifiers can be found in the volumes listing in the Bibliography and in an excellent review article (15).

REFERENCES

1. Marvin Minsky, *Proc. IRE*, **49**, 8 (1961).
2. R. J. Solomonoff, *Proc. IEEE*, **54**, 1687 (1966).
3. C. A. Rosen, *Science*, **156**, 38 (1967).
4. George Nagy, *Proc. IEEE*, **56**, 836 (1968).
5. M. D. Levine, *Proc. IEEE*, **57**, 1391 (1969).
6. L. D. Harmon, *Proc. IEEE*, **60**, 1165 (1972).
7. Kendall Preston, Jr., *Proc. IEEE*, **60**, 1216 (1972).
8. George Nagy, *Proc. IEEE*, **60**, 1177 (1972).
9. R. C. Strand, *Proc. IEEE*, **60**, 1122 (1972).
10. L. N. Kanal, *Proc. IEEE*, **60**, 1200 (1972).
11. J. R. Lox, Jr., F. M. Nolle, and R. M. Arthur, *Proc. IEEE*, **60**, 1137 (1972).
12. E. A. Patrick, F. P. Stelmack, and L. Y. L. Shen, *IEEE Trans.*, **SMC-4**, 1 (1974).
13. T. M. Cover and P. E. Hart, *IEEE Trans.*, **IT-13**, 21 (1967).
14. G. W. Gates, *IEEE Trans.*, **IT-18**, 431 (1972).
15. Y.-C. Ho and A. K. Agrawala, *Proc. IEEE*, **56**, 2101 (1968).

CHAPTER II

Introduction to Binary Pattern Classifiers

PATTERN VECTORS IN HYPERSPACE

A wide variety of types of data can be represented by points in a euclidean space of suitable dimensionality. For example, properties such as position or momentum can be specified by a point in three dimensions, or equivalently as a three-dimensional vector. The implication of this notation is that three linearly independent axes can be defined which completely define the given property in three-space when values are specified. This method of description can readily be extrapolated to more dimensions as necessary. For example, in mechanics it is common to define the position and momentum of a particle in a six-coordinate space often referred to as phase space. A single point in this six-dimensional space then simultaneously defines the position and momentum of a particle. The situation can be viewed as a vector directed from the origin to that point.

Many forms of chemical data can be represented as a d-dimensional vector:

$$\mathbf{X} = x_1, x_2, \ldots, x_d \tag{1}$$

The individual components of the pattern vector x_j are observable quantities. For example, to represent a low-resolution mass spectrum as a vector, x_j could be set equal to the intensity of the peak in the m/e position j. In the case of tabular data, a dimension can be assigned to each data column. For example, a table of melting points, boiling points, atomic weights, indices of refraction, and densities may be converted into a set of vectors in five-space, one for each compound.

9

In the case of graphical data, which is seemingly continuous, the problem of transcribing the data into the vector format is slightly more difficult. As long as data are incremental, such as those that might be represented as a bar graph, the way to assign dimensions is obvious. For example, in a low-resolution mass spectrometer with a nominal resolution of 1 mass unit and a scan from 1 to 200 mass units, a 200-dimensional vector can represent the mass spectrum of any compound. However, many devices produce analog data that are continuous. Continuous-data collection devices, such as scanning spectrometers, however, must have some resolution limit. Usually, the nature of the electronics and optics of these instruments in such that they integrate over a range, thereby defining the resolution limit. In such cases it is usually simplest to determine the resolution r and then divide the spectrum into R/r components or dimensions, where R is the scan range. Thus an infrared spectrum covering the wavelength range of 2.0 to 15.0 μm might be digitized into 130 elements if $r = 0.1$.

It is important to realize that the hyperspace notation is only one way of representing data. In general, geometric properties cannot be associated with the actual data. Thus while the orthogonal dimensions of a vector are independent, *it is certainly not the case that fragments of a mass spectrum represented in this fashion are always independent.*

Because of the orthogonal nature of the vector components, various operations may be performed on individual dimensions in a reversible fashion. Therefore an operator that treats each dimension separately may be applied without destroying the original data. This is true as long as an inverse operation exists that will reproduce the original vector. For example, various normalization procedures such as taking the square root or logarithm of each component, while changing the dynamic range of the data, are transformations for which the original vector can be generated by the inverse operation. This feature plays an important part in certain discriminant training operations.

DECISION SURFACES

One of the primary purposes of pattern recognition is the correct classification of data into categories. Any collection of data representing some fundamental entity, process, or collection therefore is referred to as a pattern. For example, the mass spectrum of a chemical compound is considered a pattern resulting from the complex chemical and physical process that produced it and, furthermore, that pattern is characterized by certain fundamental properties of the compound and the measurement process. Pattern recognition techniques are then applicable to such problems as classifying mass spectra into chemical categories. An example of such a chemical category is oxygen-containing compounds versus compounds not containing oxygen.

Representing a set of mass spectra as points in a hyperspace then yields a set of pattern points containing all the information inherent in the original spectra. The problem then becomes one of dividing the set of points into subsets as defined by the classes to be recognized, that is, how to transform pattern space into classification space.

One method of accomplishing the classification of pattern points is to find groups that belong in the same class and locate decision surfaces between them. In simple two-space this amounts to drawing lines (not necessarily straight ones) between points of different classes, for example, cows can be separated from horses by building a fence.

In hyperspace it might be expected that pattern points corresponding to compounds with similar characteristics would cluster. For example, the pattern points representing a group of mass spectra of alcohols would be expected to cluster in one limited region of the hyperspace, and pattern points representing the mass spectra of a group of alkenes would be expected to cluster elsewhere. This expectation is often met by sets of points representing chemical data such as mass spectra. When clusters occur, it is often possible to locate decision surfaces which pass between the clusters. The simplest such decision surface is a hyperplane with the same dimensionality as the hyperspace being employed.

While this hyperplane need not be linear or "flat" with respect to the dimensionality of the space, a convenient mathematical simplification occurs when the decision surface is linear and passes through the origin. Under these conditions the hyperplane can be represented by a normal vector from the origin. Or, more formally, every vector from the origin defines a plane that is the locus of points perpendicular to the vector.

Since it is very convenient to have the decision surface pass through the origin of the pattern space, it is worthwhile to ensure that this will always be possible with no loss of separability. An extra, orthogonal dimension is added to the pattern space, and all the pattern vectors are augmented with a $(d + 1)$st component. The $(d + 1)$st component of the pattern vectors can be assigned any value, but it is usually given the value of unity.

In addition to being able to represent a linear decision surface by specifying its normal vector, another important feature arises for this simple situation. The dot product of the normal vector $\mathbf{W}$ and a pattern vector $\mathbf{X}$ defines on which side of the hyperplane a given pattern point lies:

$$\mathbf{W} \cdot \mathbf{X} = |\mathbf{W}| \, |\mathbf{X}| \cos \theta \qquad (2)$$

where θ is the angle between the two vectors.

$$\cos \theta > 0 \qquad \text{for} \qquad -90° < \theta < 90°$$
$$\cos \theta < 0 \qquad \text{for} \qquad 90° < \theta < 270° \tag{3}$$

Since the normal vector is perpendicular to the plane, all patterns having dot products that are positive lie on the same side of the plane as the normal vector, and all those with negative dot products lie on the opposite side. (Points with zero dot products lie in the plane, which constitutes another definition of the location of the plane.)

While decision surfaces need not be linear, their simplicity when linear is appealing. Additionally, it can be shown that more complex decision surfaces can be implemented by linear decision surfaces preceded by appropriate preprocessing.

TLUs AS BINARY PATTERN CLASSIFIERS

The TLU is so named because it is analogous to a logical circuit element which exists in one of two states depending on whether an input is above or below a certain level (threshold).

When a binary decision is to be made, that is, when patterns are to be placed in one of two classifications, a TLU is one frequently successful method. In general some function must be used that generates one of two results based on the input.

While linear TLUs are frequently used and make good examples, TLUs of any functionality are possible. The only requirement is that a clear discrimination can be made between the two classifications of interest.

A simple bimetallic thermostat as used in a typical household heating system is a good example of a TLU. The input, room temperature, is transformed by the bending of bimetal and, as long as it is above a certain threshold, which can be defined in degrees, the thermostat puts out zero voltage and does not activate the relay to increase the furnace output. When the temperature drops below the desired threshold, a voltage is generated which can be used to activate the furnace.

By using hyperspace notation, a TLU for pattern vectors can be described as any algorithm that generates two different states from the vector as an input. While not required, zero is often used as a threshold for mathematical convenience. The linear discriminant function described earlier can be used as a TLU with great convenience. As shown above, a normal vector $\mathbf{W}$ may describe the discriminating hyperplane; the dot product of it and a pattern vector is positive for pattern vectors on the same side of the plane as the normal vector, and negative for those on the opposite side. Hence, with a threshold of zero, the linear discriminant function defines a classification space with points lying in two groups as defined by the hyperplane normal to the discriminating vector.

Note that no real restriction has been placed on the classification process by having only two results, because several TLUs may be used in conjunction to give any desired degree of discrimination. (Later examples show this process.) Coin-operated vending machines use the sizes of coins to make a series of binary decisions, for example.

Another aspect of interest in the use of linear discriminant functions as TLUs is the concept of the weight vector. A second and equivalent way to define the dot product of two vectors is:

$$\mathbf{W} \cdot \mathbf{X} = |\mathbf{W}| \, |\mathbf{X}| \, \cos \theta = w_1 x_1 + w_2 x_2 + \cdots + w_d x_d + w_{d+1} \qquad (4)$$

that is, each of the components of $\mathbf{W}$ weights each of the terms of $\mathbf{X}$. When used as a TLU, this weighting yields discrimination by causing the scalar product to fall above or below the threshold.

TRAINING OF TLUs USING ERROR CORRECTION FEEDBACK

As seen above, a TLU can be used to dichotomize a set of data represented as points or vectors in hyperspace. The problem then becomes one of finding a successful dichotomizer for a given set of classifications. This is what was meant by the earlier statement concerning the transformation of pattern space into classification space.

In many cases it is possible to classify data successfully into two classes with a linear discriminant function. However, finding such a function for multidimensional data poses an interesting problem. It is not possible simply to plot a high-dimensional graph and look at the points. Furthermore, it is frequently impractical to calculate all possible decision surfaces in order to find a successful one. However, a heuristic method has proven quite successful in many cases.

This heuristic approach is based on selecting a starting classification surface or discriminant function, either arbitrarily or by some approximation scheme, and then "training" the classifier by modifying it as it accumulates experience in making its decisions. This training is done with a set of patterns for which the classifications are known (the training set). The patterns are presented to the classifier being trained one at a time and, when incorrect classifications are made, the decision surface is altered in order to correct the error.

There are several schemes that are called error correction feedback. One method, which can be shown in converge for any linearly separable set of data, corrects the decision plane by reflecting it about the misclassified point. As described earlier, the dot product of the weight vector and a pattern vector gives a scalar whose sign indicates on which side of the decision surface the pattern point lies:

$$\mathbf{W} \cdot \mathbf{X} = s \tag{5}$$

(An arbitrary decision must be made as to which subset of the data is to be called the positive class and which is to be called the negative class.) When pattern i of the training set is misclassified, then

$$\mathbf{W} \cdot \mathbf{X}_i = s \tag{6}$$

in which s has the incorrect sign for classifying $\mathbf{X}_i$. The object is to calculate an improved weight vector $\mathbf{W}'$, such that

$$\mathbf{W}' \cdot \mathbf{X}_i = s' \tag{7}$$

where the sign of the scalar result s' is opposite what it was previously. The new weight vector is calculated from the old one by adding an appropriate multiple of $\mathbf{X}_i$ to it:

$$\mathbf{W}' = \mathbf{W} + c\mathbf{X}_i \tag{8}$$

Combining equations 7 and 8 gives

$$s' = \mathbf{W}' \cdot \mathbf{X}_i = (\mathbf{W} + c\mathbf{X}_i)\mathbf{X}_i \tag{9}$$

which can be algebraically rearranged to give

$$c = \frac{s' - s}{\mathbf{X}_i \cdot \mathbf{X}_i} \tag{10}$$

It remains only to choose a value for s' to complete the derivation. An effective method is to let $s' = -s$. This moves the decision surface, so that after the feedback correction the point $\mathbf{X}_i$ is the same distance on the correct side of the decision surface as it was previously on the incorrect side. If $s' = -s$ is put into equation 10, then

$$c = \frac{-2s}{\mathbf{X}_i \cdot \mathbf{X}_i} \tag{11}$$

and $\mathbf{W}'$ can be calculated directly by using equations 11 and 8.

The training procedure involves iterating over all the pattern points in the training set and correcting the weight vector whenever an error is committed, until the discriminant function converges on one that successfully classifies all the points. The process is completely analogous to asking repeatedly a series of questions until the device being trained responds correctly to all of them. It is this type of approach that had led to the use of such terms as "learning" and "learning machine," that is, the definition of learning used in this sense is the improvement of performance with experience.

As mentioned above, this error correction procedure can be shown to find a solution if one exists. Therefore the weight vector can be initialized arbitrarily, although it is obviously better practice to use whatever information is available to estimate a starting weight vector.

A word of caution is in order with respect to the question of the training set size N versus the number of components per pattern d. While the exact ratio necessary for obtaining valid results is in question, it is generally agreed that the ratio N/d should be as large is possible. An added problem is that the value of d is not so important as the number of descriptors in the patterns necessary for separation of the two classes. This number is not usually known a priori. It is now generally accepted that, if N/d is greater than approximately 3, the results obtained will not be in question.

PROPERTIES OF TLUs

Four properties are discussed with respect to TLUs and other pattern classification devices. These are *recognition, convergence rate, reliability*, and *prediction*.

Recognition is defined as the ability of a discriminant function to correctly classify those patterns with which it was developed, that is, its training set. How well can it answer questions that were used to develop it? This is analogous to asking on examinations virtually the same questions that were covered in class; however, it may have great utility, by reducing the classification procedure to a simple mathematical operation rather than requiring the retention and retrieval of a large library of data. This may be much more economical in many applications. In this book 100% recognition is analogous to perfect or error-free training of a classifier.

Convergence rate refers to the speed with which a training algorithm converges toward 100% recognition. This is of interest with regard to the economics of developing useful pattern classifiers. Rapid convergence is useful to those who have limited computer budgets.

While slow convergence does not mean a problem is insolvable, it often means that it cannot be done economically. Therefore in many cases it may be necessary to trade or compromise other advantages of classifiers for improved convergence rate.

Reliability refers to the ability of a classifier to classify correctly data which were used in its development but which are classified after undergoing some distortion. All processes of information transfer have some noise level. For example, while a classifier might have been developed with mass spectra from a certain standard set of chemical compounds, repeated runs of the same compound do not produce exactly the same mass spectrum. The degree of reliability indicates the classifier's ability to handle such distorted or noisy data. Furthermore, it is related to the redundancy of the decision procedure itself. Certainly, human pattern recognition systems have incredibly high reliability for certain classifications. Only a small subset of a large body of possible data is necessary for a human beings to recognize pussycats optically. Such fundamental properties as size, color, weight, and others, have little to do with the decision process. Rather, general features of shape, type of movement, and so on, are more important. It is certainly important for the application of pattern recognition to chemical problems that noise levels may be tolerated. Furthermore, this is one of the inherent advantages that pattern recognition techniques may have over more conventional direct comparisons of data with library compilations.

Prediction is probably the most exciting aspect of pattern recognition applications to chemistry. Prediction refers to the ability of the classifier to classify correctly patterns which were not members of the training set. This amounts to the "unknown" test. If a pattern classifier can be shown to predict unknown data successfully, then it is implied that some of the fundamental relations between data and their classifications have been extracted in the development of the discriminant function. Prediction may not only answer questions on the unknown data, but the way it is accomplished may suggest fundamental relations and lead to further understanding of cause-and-effect relations in chemistry.

CHAPTER III

Preprocessing
and Transformations

The success of pattern recognition techniques can frequently be enhanced and/or simplified by prior treatment of the data before the classification stage. In terms of the n-dimensional euclidean space, called the pattern space, being used, the goals of preprocessing or transformation are:

1. To separate clusters of related points further apart to make classification easier
2. To reduce the dimensionality of the pattern space to make classification more economical

These two goals are often contradictory. However, if separation between classes is improved, classification becomes easier. Furthermore, reducing the dimensionality of data makes classification more economical, since the time required for calculation of the discriminant function is frequently proportional to the dimensionality of the data. Also, reduction of dimensionality can reduce the size of the data set required to avoid underdetermined situations.

Several terms frequently used in conjunction with the pretreatment of data are: preprocessing, transgeneration, or transformation; feature selection; and feature extraction. *Preprocessing* and *transgeneration* refer to altering the individual components of pattern vectors. While various functionalities may be applied, this constitutes what we call a dimensionally independent transformation. *Feature selection* refers to selecting those features of the raw data believed to be more important. *Feature extraction* refers to combining individual primitive features into higher-level ones. This involves transformations that do not preserve dimensional independence, such as for the formation of cross terms, Fourier transformations, and processes such as factor analysis.

This chapter is devoted to several examples of dimensionally independent linear preprocessing of chemical data. Since the preprocessing stage cannot be separated from the remainder of the pattern classification system, overall performance of a binary pattern classifier must often be used to compare different preprocessing approaches. Accordingly, this chapter provides some figures of performance of entire pattern recognition systems; however, this practice is kept to a minimum, since most of this type of information is presented in subsequent chapters. The first sections of this chapter deal with preprocessing of single-source data in which the measurements are intrinsically linked. Later discussion focuses on the preprocessing of multisource or diverse data when the various data are from unrelated measurements.

MASS SPECTRA

The first type of chemical data to be discussed are mass spectra. A study in which a comparison of several methods of preprocessing were applied to low-resolution mass spectra is described in the following discussion (1).

The work used a data set taken from a collection of mass spectra purchased on magnetic tape from the Mass Spectrometry Data Center, Atomic Weapons Research Establishment, United Kingdom Atomic Energy Authority, Aldermaston, Berkshire. A portion of that tape contains 2261 American Petroleum Institute (API) Research Project 44 spectra. Six hundred of these spectra were digitized with normalized intensities ranging from 99.99 to 0.01 in each spectrum. Spectra corresponding to compounds containing only carbon, hydrogen, oxygen, and nitrogen atoms were used. In the data selected from the tape for actual tests, there were about 50 to 70 such peaks per spectrum. There were 132 m/e positions in which 10 or more peaks appeared throughout the data set used. In making actual computer runs, an input subroutine was employed to read spectra from a tape, select the ones to be used according to preset criteria, for example, carbon number, and set up the data for the remainder of the program to use. Normally, 600 spectra were thus input, with a training set of 300 and a prediction set of 300.

Because of the large quantity of data available, it was possible to select data for inclusion in each problem so as to acquire as homogeneous a data set as possible. In the present example the spectra were selected to correspond to compounds with 3 to 10 carbon atoms. (Tests using data sets with spectra of compounds with 3 to 20 carbon atoms showed that the results obtained in this investigation were not artifacts of the data, but worked for the less homogeneous data set also.) Six hundred spectra meeting the criteria of 3 to 10 carbon atoms were used for each computer run; they were divided evenly between the training and prediction sets. The data set of 600 spectra most often used contained 35,550 peaks spread over 132 m/e positions.

The mass spectra were transformed by four dimensionally independent methods: square root, fourth root, logarithmic transform, and zeroth power transformation. The zeroth power transform actually amounts to one-bit encoding, that is, where a peak occurs in the spectrum that peak is defined as a 1, and where no peak occurs it is defined as a 0. This could also be viewed as a thresholding process. The results obtained are shown in Table 1.

For each of the four transformations, a feature selection program was run. The main points of the feature selection routine were that for each stage in the routine two weight vectors (initiated with all the components set equal to $+1$ and all equal to -1) were trained, and then the signs of the weight vector components compared. Only the m/e positions corresponding to weight vector components with the same sign were retained for the next stage. The cycle was repeated until no more ambiguous m/e positions could be located, and the program was then terminated. Each row of Table 1 shows the results of running this program for the same data, having undergone the transformation listed in column 1. The original data have a dynamic range within each spectrum of 10^4 (0.01 to 99.99); the second column gives the dynamic range after the transformation. The third column lists the number of feedbacks necessary for convergence to 100% recognition for the first stage of the feature selection program (when there are 132 m/e positions). The fourth column gives the predictive abilities of the two pattern classifiers, and the fifth lists the average percent prediction for this first stage. The logarithmic transformation yields the best predictive ability for the first step of the program. Column six shows the number of ambiguous m/e positions found during this first stage; a narrower dynamic range causes the pattern classifier to find few ambiguous m/e positions. The program repeats the cycle mentioned above many times for each transformation, and the number of m/e positions remaining after all ambiguous positions have been discarded is shown in column 7. Again, narrower dynamic range reduces the number of ambiguous m/e positions. The final column gives the predictive ability exhibited by each pattern classifier averaged over all stages of the feature selection process. Once again the logarithmic transformation exhibits the highest predictive ability. The result is consistent with the result obtained by applying information theory to the transformation problem. Throughout the remainder of the original study, the logarithmic transformation was employed.

TABLE 1 Effects of Transformations on Properties of Binary Pattern Classifiers

Transformation	Dynamic range	Feedbacks +1 WV/−1 WV	Per cent prediction +WV/−WV	Av % prediction	m/e positions discarded	m/e positions final	Overall % prediction
Square root	100	123/114	93.7/96.3	95.0	66	43	94.4
Fourth root	10	101/107	94.7/94.3	94.5	49	44	94.6
log	4	141/102	95.3/95.3	95.3	44	57	95.5
Zeroth power	1	235/229	93.7/94.0	93.8	23	83	94.0

	+	−	Total
Training set	121	179	300
Prediction set	126	174	300

132 m/e positions
Oxygen presence–absence determinations

An information theory argument has been presented (2) in favor of the logarithmic transgeneration function. In this work, involving the compression of mass spectra, several transition levels of intensity were set logarithmically at 0.5, 1, 2, 4, 8, 16, and 32% of the total ion current. This was an attempt to guarantee roughly equal numbers of peaks at each level, thereby maximizing the information content or the so-called independent channel information entropy.

The argument addressed to the dynamic range of the variables proves to have a major effect on the convergence of TLUs. It has been demonstrated in several studies that making dynamic ranges approximately equal for all descriptors often gives the best performance. Since this is a dimensionally independent transformation, it has no effect on the separability of a binary data set. However, any process that makes convergence quicker saves computer time and is obviously of value.

Reference 3 showed that the dynamic range of various components in the patterns can vary considerably, but beyond certain limits large dynamic ranges cause a great increase in convergence time. In this particular study the value assigned to the $(d + 1)$st component of each pattern vector was varied over several orders of magnitude; it was demonstrated that, when it was made much larger than the average values of the other components, convergence became very slow.

Several works (e.g., 1, 4, 5) have shown that binary mass spectra have a great deal of information content, and in many cases are perfectly suitable for pattern recognition applications.

INFRARED SPECTRA

The same arguments used above with reference to the preprocessing of mass spectra data can also be used with reference to infrared spectra. In reference 6 infrared spectra are represented with only four amplitude values.

The data used in this study were made available on loan from the Sadtler Research Laboratories, Inc., Philadelphia, Pa., and were part of the same data used in the Sadtler IR Prism Retrieval System. From 24,142 spectra of standard compounds, the first 4500 were selected that satisfied the requirements of no more than 10 carbon atoms, 4 oxygen atoms, 3 nitrogen atoms, and no additional elements except hydrogen. These Sadtler spectra were recorded in 0.1-μm bands from 2.0 to 14.9 μm. This gave a total of 130 pattern dimensions, including the $(d + 1)$st term. The amplitude of each pattern component was assigned one of four values based on the intensity of the strongest absorption in that 0.1-μm band. For the largest peak in the spectrum, the amplitude was set at 3.0; for the largest peak in a 1.0-μm band, the amplitude was 2.0; for other peaks, the amplitude was 1.0; and for no peak in a given 0.1-μm band, the amplitude was set at 0.0. Because the amplitude data were limited to only three nonzero values, and the majority of the values were zero, it was useful to compress the patterns to save computer storage and decrease computation time. Each pattern was therefore rewritten as a series of integers:

$$n_1, p_1, p_2, p_3, \ldots, p_{n1}, n_2, q_1, q_2, q_3, \ldots, q_{n2}, n_3, r_1, r_2, r_3, \ldots, r_{n3}$$

where n_1 is the number of peaks with an amplitude of 1.0 followed by the dimensions (or positions) of those peaks, n_2 is the number of peaks with an amplitude of 2.0 followed by the dimensions of those peaks, and n_3 is the number of peaks with an amplitude of 3.0 followed by the dimensions of those peaks. The resultant patterns required less than one-third the storage of the compressed spectra. Furthermore, the dot product process used to form the scalars necessary for classification can be performed as follows. If $\mathbf{W}$ is the weight vector and $\mathbf{Y}$ is the pattern for which the dot product is to be formed, then

$$\mathbf{W} \cdot \mathbf{Y} = w_1 \cdot y_1 + w_2 \cdot y_2 + w_3 \cdot y_3 + \cdots + w_{d+1} \cdot y_{d+1} \tag{1}$$

However, if $\mathbf{Y}$ contains only values of 0.0, 1.0, 2.0, and 3.0, the dot product may be computed by

$$\mathbf{W} \cdot \mathbf{Y} = \left(\sum_{j=1}^{n_1} w_{pj} \right) + 2.0 \left(\sum_{j=1}^{n_2} w_{qj} \right) + 3.0 \left(\sum_{j=1}^{n_i} w_{rj} \right) + w_{d+1} \tag{2}$$

For the patterns used, this method of computing dot products decreased computation time by roughly a factor of 20.

An independent study of infrared spectra interpretation used a rather different data representation (7). The data were taken from the Sadtler Standard Infrared Spectra. Each spectrum was divided into 0.1-μm intervals across the 2.0 to 15.0 μm region to give 131 intervals and therefore 131 xy values. The transmittance in each interval was converted into absorbance and put on a 0 to 9 scale for convenience. Thus each infrared spectrum was represented by a 131-dimensional pattern vector in which each component was an integer between 0 and 9.

MULTISOURCE SPECTROSCOPIC DATA

Another convenient aspect of pattern recognition is the ability to take data from diverse sources and treat it as a single pattern. For example, in the linear learning machine, individual spectral components are treated independently. The result is as if the spectra of 100 abscissa positions were really

100 independent experiments. Hence data from radically different sources, such as infrared and mass spectra of the same compound, may be combined and treated as a single pattern. After all, these two spectroscopic measurements are both attempts to elucidate information on the same compound and, while the data may be collected in radically different ways, arguments could be made that the relation between certain components of the fragmentation pattern and the infrared adsorption are no more unrelated than certain combinations of infrared absorptions themselves.

Reference 5 describes an investigation of the combination of infrared spectra, mass spectra, and melting and boiling points as a pattern representative of a chemical compound. Here the dynamic range argument became very important, because of the arbitrary scaling of the data from the different techniques. When the magnitude of the values from one technique was much larger than that of the other, the former was predominant in the pattern recognition result. However, when the two techniques were normalized to the same dynamic range, pattern features could be extracted from both sources and maximum prediction resulted.

Patterns were generated for 291 compounds for which both the low-resolution mass spectrum and the infrared spectrum were available. The mass spectra, which were from the API Research Project 44 tables, had a total of 132 possible mass positions and had amplitudes adjusted in a range from 10 to 100 by square-root transformation. The infrared spectra were obtained on loan from Sadtler Research Laboratories, Inc., and contained 130 possible absorption wavelengths, over a range of 2.0 to 14.9 μm. Four peak amplitudes are used in the infrared data: (1) peak not present, normally assigned 0.0 intensity; (2) peak present in 0.1-μm band, normally given intensity of 1.0; (3) largest peak in a 1.0-μm band, normally given an intensity of 2.0; and (4) largest peak in the spectrum, given an intensity of 3.0. Thus an infrared spectrum consisted of an ordered sequence of 130 numbers. Compounds had formulas of $C_{1-10}H_{1-24}O_{0-4}N_{0-3}$, and in each of the combined patterns of the mass spectrum and infrared spectrum there were 262 components. Of the 291 such patterns, 191 were selected randomly as a training set, and the other 100 were used as a predicting set. In combining data from different sources into a single pattern, the relative contributions of the two (or more) types of data grossly affected the learning machine's behavior.

In order to evaluate the effect of combining data from diverse sources, compounds were classified based on the presence or absence of one or more double bonds. Attempts to answer this question using only infrared patterns or mass spectrometry patterns met with limited success, as shown in the first two sections of Table 2. In each case parameters were discarded from

TABLE 2 Double-Bond Presence

1. IR   2. MS

Parameters	No. of feedbacks	Recognition	Per cent prediction	No. of feedbacks	Recognition	Per cent prediction
262						
162						
125	156	x	79	1105	x	87
100	144	x	83	958	x	88
70	177	x	82	1529	x	85
50	261	x	77	1677	x	84
30	>5000	168	62	>5000	179	81
20	>5000	136	47	>5000	178	81
10	>5000	129	50	>5000	141	69
5	>5000	129	69	>5000	131	56

3. Combined patterns

Parameters	No. of feedbacks	Recognition	Per cent prediction	MS/IR
262	1182	x	86	136/126
162	1024	x	84	99/6
125	1005	x	87	92/33
100	1056	x	85	88/12
70	1176	x	85	69/1
50	1845	x	85	50/0
30	>5000	182	79	30/0
20	>5000	179	82	20/0
10	>5000	135	61	10/0
5	>5000	142	61	5/0

24

5. Combined patterns

No. of feedbacks	Recognition	Per cent prediction	MS/IR
105	x	89	136/126
82	x	88	76/86
82	x	89	61/64
87	x	90	46/54
93	x	89	30/40
169	x	89	23/27
146	x	92	13/17
>5000	174	88	9/11
>5000	164	75	6/4
>5000	137	62	4/1

4. Combined patterns

Parameters	No. of feedbacks	Recognition	Per cent prediction	MS/IR
262	141	x	78	136/126
162	149	x	80	52/110
125	145	x	83	24/101
100	133	x	80	11/89
70	121	x	81	1/69
50	128	x	78	0/50
30	>5000	186	69	0/30
20	>5000	140	52	0/20
10	>5000	136	52	0/10
5	>5000	139	55	0/5

6. Combined patterns

Parameters	No. of feedbacks	Recognition	Per cent prediction	MS/IR/other
262	87	x	87	136/126/2
162	83	x	88	76/76/2
125	90	x	89	55/70/2
100	88	x	89	44/56/2
70	121	x	91	31/39/2
50	129	x	90	25/25/2
30	468	x	89	15/15/2
20	1520	x	92	12/8/2
10	>5000	176	84	8/2/2
5	>5000	129	64	5/1/1

the 125 starting parameters by a feature selection routine which assumed that pattern vector components that have small corresponding weight vector components are relatively unimportant. Prediction is about 82% for the infrared case, and falls off rapidly when the number of dimensions is below 50. Mass spectrometry gives prediction in the region of 87%, which slowly decreases until the number of parameters is reduced below 20, when it drops off rapidly toward random success. The mass spectrometry case also fails to converge within the allotted number of feedbacks below 50 parameters.

Sections 3 to 5 of Table 2 show the results of training with patterns formed by combining the mass spectrum and infrared spectrum. In section 3 the patterns have been normalized such that the mass spectral peak intensities are much greater than the infrared components. The mass spectral peak intensities were in the range 10 to 100, and the infrared intensities were set to 0, 1.0, 2.0, 3.0. Thus the contributions from the infrared data to the overall length of any pattern vector are relatively small compared to that from the mass spectrum portion of the pattern. As is expected, the results are almost identical to those of the mass spectra alone and, by the time the number of parameters reaches 50, there remain only components of the mass spectral patterns. In section 4 the results of normalizing the patterns so that the infrared data predominate is shown. Here the infrared intensities are set to 0, 1000, 2000, 3000, so most of the pattern vectors' length is due to the infrared contribution. As expected, the results are quite like the independent infrared spectra, and when only 50 parameters are left, they are all from infrared patterns.

Section 5 shows the results of normalizing the patterns so that each data source contributes equally to the total amplitude of the pattern set. The infrared intensities over the entire data set are equal to the sum of all the mass spectral peak intensities over the entire data set. In this case the prediction starts out at about 90% and remains very high until the number of pattern components is reduced below 20. Even with only 10 components left, prediction is 75% and recognition is 82%, showing better than random behavior. It is also interesting to note that components of both the mass spectra and the infrared spectra have been retained throughout the parameter reduction process.

Section 6 of Table 2 shows the further improvement gained by adding two more data points in the patterns, the melting point and boiling point for each compound. (Note that, in every case in which the boiling and melting points are included, the total number of parameters is two more than in the combined mass spectrum–infrared spectrum patterns of the uncombined patterns.) In comparison to section 5, no particular improvement is noted down to 30 parameters, both the predictive abilities and convergence rates

TABLE 3 *Double-Bond Presence with Binary Spectra*

	1. IR			2. MS		
Parameters	No. of feedbacks	Recognition	Per cent prediction	No. of feedbacks	Recognition	Per cent prediction
262						
162						
125	183	x	81	454	x	84
100	173	x	83	486	x	85
70	202	x	79	373	x	85
50	193	x	79	494	x	84
30	>5000	172	77	>5000	175	85
20	>5000	165	65	>5000	161	81
10	>5000	144	56	>5000	157	75
5	>5000	142	61	>5000	155	69

3. Combined patterns

Parameters	No. of feedbacks	Recognition	Per cent prediction
262	124	x	92
162	110	x	89
125	122	x	90
100	119	x	89
70	99	x	88
50	105	x	89
30	234	x	89
20	>5000	178	85
10	>5000	161	54
5	>5000	158	72

27

being about the same. However, with addition of the boiling and melting points, the learning machine still converges with only 20 parameters, and also retains approximately a 90% predictive ability at that level. Furthermore, the prediction at 10 parameters is still notably higher than in the other cases. The decision process used to reduce the number of parameters contributing little to the classifications retains the melting and boiling point information almost until the end of the calculation.

In much analytical instrumentation, such as infrared and mass spectroscopy, it is often easier to obtain accurate information on the resolved property in the measurement—for example, wavelength in the former case and mass value in the latter— than on the intensity of the measurement. For this reason it is useful to be able to evaluate how important intensity information really is. Table 3 shows such an evaluation for the double-bond determination summarized in Table 2. In this case all components of both the mass spectrum and the infrared spectrum were reduced to intensity values of 1.0 where detectable peaks occurred, and 0 where they did not. These "binary spectra" were then trained in the same fashion as before. The results of sections 1 to 3 of Table 3, when compared to sections 1, 2, and 5 of Table 2, show that apparently the peak–no peak information is just as good as the intensity-resolved information for this question. In other words, information sufficient to answer the question of double-bond presence is present in peak locations for both infrared and mass spectrometry. Hence, in this particular case, it is only necessary to have the peak positions to train and predict concerning the question of double-bond presence.

Another investigation involving the use of diverse data is reported in reference 8. The data used included low-resolution mass spectra, nuclear magnetic resonance (NMR) data, refractive indices, and densities of pure hydrocarbon compounds. These data were used in an attempt to determine the hydrocarbon types and the structure of the average molecule in a complex mixture of hydrocarbons, namely, a gasoline sample. These pattern vectors formed from diverse data were inserted into a pattern recognition algorithm using a least-squares training procedure.

ELECTROCHEMICAL SPECTRA

In reference 9 stationary electrode polarograms (SEPs) were investigated using pattern recognition techniques. The object of the investigation was to determine the utility of the pattern recognition technique for distinguishing SEPs due to only one chemical species from those due to two or more species. As in any pattern recognition investigation, the first step was to develop several features of the data to be used in constructing pattern vectors.

A set of SEP curves was developed from analytical functions, and these data were digitized and stored. The data were then treated as if they were experimental data. A set of 133 features was derived from the digitized current-versus-time SEP curves and their first and second derivatives (Figure 1). Features used included such measures as: change in current with a change in potential at specific points before and after the peak of the current curve; changes in the potential at various fractions of the maximum current; measures of skewness; areas under the various portions of the plots and ratios among these areas, and various cross-derivative parameters. These features were assigned the general names shape features, peak current features, and peak potential features.

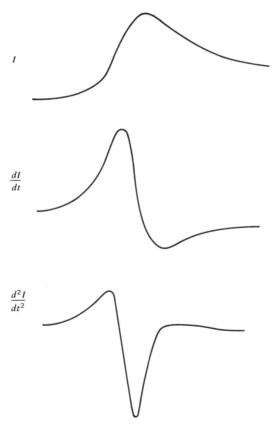

Figure 1.   *Stationary electrode polarogram plots.*

By the nature of the feature development process, it is seen that each feature has its own range and distribution of values. This necessitates performing some preprocessing to alleviate any potential problems.

Two methods of transforming the data were employed. The first was called range fixing. In this procedure the value of the ith feature for the jth SEP is adjusted from its original value x_{ij} to an adjusted value x'_{ij} by the equation

$$x'_{ij} = a_i x_{ij} + b_i \tag{3}$$

in which a_i and b_i are constants related to the maximum and minimum values of the feature under investigation. These parameters can be chosen in order to fix an arbitrary range for feature i.

The second method used for transforming the data was autoscaling. Autoscaling has also been discussed in reference 10. Here

$$x'_{ij} = \frac{x_{ij} - \bar{x}_i}{\sigma_i} \tag{4}$$

where $\bar{x}_i$ is the mean and σ_i the standard deviation of the values x_{ij} for the ith feature. Autoscaling yields a set of values for the ith feature which have a mean of zero and unit standard deviation. The authors used data that had been transformed with the autocorrelation function throughout their work. The results obtained with these data are shown in Chapter IV.

REFERENCES

1. P. C. Jurs, *Anal. Chem.*, **43**, 22 (1971).
2. S. L. Grotch, *Anal. Chem.*, **43**, 1362 (1971).
3. L. E. Wangen and T. L. Isenhour, *Anal. Chem.*, **42**, 737 (1970).
4. P. C. Jurs, B. R. Kowalski, T. L. Isenhour, and C. N. Reilley, *Anal. Chem.*, **41**, 690 (1969).
5. P. C. Jurs, B. R. Kowalski, T. L. Isenhour, and C. N. Reilley, *Anal. Chem.*, **41**, 1949 (1969).
6. B. R. Kowalski, P. C. Jurs, T. L. Isenhour, and C. N. Reilley, *Anal. Chem.*, **41**, 1945 (1969).
7. R. W. Liddell, III, and P. C. Jurs, *Appl. Spectros.* **27**, 371 (1973).
8. D. D. Tunnicliff and P. A. Wadsworth, *Anal. Chem.*, **45**, 12 (1973).
9. L. B. Sybrandt and S. P. Perone, *Anal. Chem.*, **44**, 2331 (1972).
10. B. R. Kowalski and C. F. Bender, *J. Amer. Chem. Soc.*, **94**, 5632 (1972).

CHAPTER IV

Discriminant Function Development

While many discriminant functions are possible, the simplest to implement, and therefore the most popular in chemistry, is the linear discriminant function. As discussed earlier, a linear discriminant function amounts to a weighting function which when multiplied by a pattern vector gives a scalar result. While multicategory classification is possible, the simplest possible classifier is the binary type which gives one of two results. In the case of a linear discriminant function applied for binary pattern classification, the sign of the scalar can be very conveniently used to indicate the classification.

Most of this chapter is devoted to results obtained by applying linear binary pattern classifiers to chemical data interpretation problems. Additional sections discuss nonlinear classifications and multicategory classifications.

BINARY PATTERN CLASSIFIERS

Elementary TLU

As shown above, the basic TLU makes classifications by calculating the dot product between the weight vector $\mathbf{W}$ and the pattern vector $\mathbf{X}$ being classified. The sign of the scalar result is used for classification:

$$\text{If } s > 0, \text{ then } \mathbf{X} \text{ is in category 1}$$

$$\text{If } s \leq 0, \text{ then } \mathbf{X} \text{ is in category 2}$$

The weight vector is found by using an error correction feedback procedure in conjunction with a training set of known pattern vectors. This section is concerned with studies of this most basic implementation of a TLU.

An early investigation of convergence rates and predictive ability of TLUs is described in reference 1. The data are from the API research Project 44 tables. The overall data set consisted of 630 low-resolution mass spectra of compounds with molecular formulas $C_{1-10}H_{2-22}O_{0-4}N_{0-2}$. The intensities of the peaks were reported as a percentage of the base peak in each spectrum. Only peaks with intensities greater than 1% of the base peak were used; most of the spectra have 15 to 40 such peaks. Within the entire data set there were peaks in 155 m/e positions, so the patterns were 156-dimensional vectors. The intensities put into the pattern vectors were subjected to pre-processing consisting of taking the square root of the reported intensity. Table 1 shows the results of training weight vectors to determine oxygen presence, using different-sized training sets chosen randomly from an overall data set of 630 low-resolution mass spectra. Normally, the convergence rate decreases as the training set becomes larger. This trend is seen in Table 1 where all weight vectors were trained to complete recognition. It is interesting to note that predictive ability is high, even for training set sizes of 50 and 100, where there are more adjustable parameters (weight vector components corresponding to mass positions) than there are patterns in the training set. Predictive ability is expected to increase as the training set size increases, and Table 1 demonstrates this, although the noise level is considerable.

The increase in predictive ability of any classifier as a function of training set size is indicative of a very fundamental point of importance in all forms

TABLE 1   Convergence Rate and Prediction of Oxygen Presence as a Function of Training Set Size

Training set size	Spectra tested*	Per cent predicted	Average per cent predicted
50	226	88·3	
	180	84·3	86·2
	255	85·9	
100	654	90·8	
	561	87·7	88·2
	574	86·0	
200	1528	86·6	
	1525	88·6	88·4
	2189	88·1	
300	2590	92·7	
	1959	90·6	90·6
	2986	88·5	

* Number of spectra required to produce complete training.

of pattern recognition. As Chapter 1 indicated, Bayes' decision theory allows the computation of the most probable class for a pattern when the distribution of the patterns is known. Unfortunately, the distribution of the universal set of a given type of pattern is often not known. For example, in the case of chemical structure analysis, it is estimated that a few million chemical compounds have been synthesized. For these, a reasonably large subset of mass spectra have been obtained. However, most consistent mass spectral compilations include spectra from only a few thousand compounds. And even if all the spectra had been measured under standard conditions for all known chemical compounds, this would still only represent a small subset of all possible chemical compounds. Hence any classification scheme must be based on an estimate of the relevant probability distribution unless a well-established theory exists to predict the mass spectra of any possible chemical compound.

Therefore whatever pattern recognition scheme is used must in itself estimate in some form the distribution of unknown mass spectra from a set of known spectra. The linear binary pattern classifier accomplishes this by finding a linear decision surface which successfully divides the multidimensional space used to represent the mass spectra into the two classes of interest. This classifier is then used to predict the classifications of unknown spectra. While this treatment is couched in a form of mathematics not normally encountered by the chemist, it amounts to no more than using the results of past experience to predict new answers. Hence the training and prediction procedure is a very common method by which scientists seek to interpret many types of data. An advantage of this particular method, when formulated as in the linear binary pattern classifiers, is that it can be tested. This is done by using one randomly selected subset of the known data set for a training set, and then using the remaining data for testing after training has been completed. The results of this testing are a measure of the predictive ability, since the data used for testing were not used during training. While there is always the chance that an artifact may produce unreasonably good predictive ability, still there is a strong probability that a binary pattern classifier with a high predictive ability indeed does relate to the fundamental phenomenon measured.

Reference 2 reports an early attempt to test the extent to which binary pattern classifiers could be used to determine molecular structure parameters from low-resolution mass spectrometry data. The same data set as described immediately above was used. However, for this investigation it was divided into two subsets: 387 hydrocarbon (CH) spectra and 243 spectra from compounds containing oxygen and/or nitrogen (CHON).

From the 387 CH compounds, a subset of 200 was chosen randomly to

TABLE 2 CH Class

		Training set			Prediction set		
	Cutoff	Negative category	Positive category	Spectra feedback	Negative category	Positive category	Per cent prediction
Carbon number	9	163	37	227	154	33	89.3
	8	121	79	167	113	74	92.5
	7	80	120	185	77	110	93.6
	6	53	147	99	44	143	94.1
	5	30	170	71	21	166	97.9
	4	15	185	42	7	180	97.3
Hydrogen number	20	196	4	53	182	5	97.3
	18	168	32	170	154	33	95.7
	16	143	57	202	132	55	97.3
	14	110	90	58	100	87	94.1
	12	72	138	51	55	132	95.2
	10	47	153	59	34	153	96.8
	8	28	172	31	19	168	96.8
	6	9	191	34	10	177	97.3
Carbon:hydrogen ratio	$2n + 2$	156	44	25	143	44	96.8
	$2n$	125	75	28	107	80	96.8
	$2n - 2$	153	47	36	154	33	96.8
	$2n - 4$	191	9	39	180	7	98.9
	$2n - 6$	185	15	13	170	17	98.9

Fragment	Count						
Methyl	4	191	9	177	165	22	90.9
	3	160	40	800	136	51	86.1
	2	114	86	859	97	90	86.6
	1	62	138	648	47	140	86.6
	0	29	171	328	24	163	89.3
Ethyl	1	166	34	1518	153	34	80.7
	0	104	96	>2000	97	90	73.3
n-Propyl	1	191	9	211	177	10	90.4
	0	145	55	>2000	148	39	71.7
Largest ring	6	199	1	11	184	3	97.9
	5	142	58	141	137	50	89.8
	4	121	79	149	114	73	90.9
	3	120	80	194	112	75	91.4
Branch point carbons	2	174	26	356	163	24	90.9
	1	108	92	>2000	90	97	61.5
	0	42	158	1486	36	151	87.2
Number of —C=C—	2	171	29	11	165	22	98.4
	1	159	41	153	158	29	92.5
	0	102	98	>2000	98	89	77.5
Carbons w/o hydrogens	1	167	33	432	156	31	83.4
	0	111	89	1327	97	90	70.6
Benzene ring	0	179	21	37	168	19	96.8
—C≡C—	0	184	16	163	174	13	93.6
Vinyl	0	166	34	>2000	158	29	80.2

serve as a training set for developing weight vectors, and the other 187 were used to test the predictive ability of the weight vectors. Table 2 shows the results of training and testing the 43 weight vectors developed for the determination of structural parameters of hydrocarbons. Each vector was trained to give a binary decision. In all cases, except the carbon/hydrogen ratio, a positive answer indicated the value was greater than the cutoff number, and a negative value indicated that the value was less than or equal to the cutoff. For example, the first vector (carbon number 9) was trained to give a positive dot product with a mass spectrum of a compound containing 10 carbons, and a negative dot product with one containing 9 or less carbons. The carbon/hydrogen ratio was trained to give a positive result for that particular ratio and a negative result for any other. For example, n-hexane gives a positive dot product for carbon/hydrogen ratio $2n + 2$ and a negative dot product for all other ratios. The categories appearing in Table 2 are defined as follows. Methyl, ethyl, and n-propyl numbers are the number of each group that can be produced by a single bond rupture, that is, 3-methylhexane has three methyls, two ethyls, and one n-propyl by this definition. The largest ring classification includes saturated, unsaturated, or aromatic rings. Branch-point carbon number is the number of carbon atoms in the compound that are bonded directly to at least three other carbon atoms. For number of carbon–carbon double bonds, benzene has been classed as three. The carbon w/o hydrogen category refers to carbons that are not bonded to any hydrogens. The final three weight vectors detect the presence or absence of benzene rings, acetylenic bonds, and vinyl structural features.

The third and fourth columns of Table 2 indicate the number of cmpounds of the training set that fell into each class. The fifth column gives the number of feedbacks necessary for convergence, with >2000 indicated for those that had not been completely trained after 2000 feedbacks. The sixth and seventh columns indicate the number of compounds in the prediction set that fell into each class. The final column gives the prediction success for the 187 compounds that were not part of the training set. Predictive ability ranged from 61.5 to 98.9%, with an average of 90.3%. The prediction percentage can be used as a gauge of the credibility of an answer when produced for an unknown spectrum. Random guessing gives 50% success. It is seen that in the CH class considerable structural information may be derived with a high-confidence level from a completely empirical calculation method.

From the 243 CHON compounds, a subset of 150 was chosen for training, and the other 93 were used to test the predictive ability of the developed weight vectors. Table 3 shows the results of training the 65 weight vectors developed for various aspects of structure. The categories appearing in Table 3 are defined as follows. The clump number of a compound is the largest number aggregate of carbon atoms bonded together. Methyl, ethyl, largest ring, and numer of double bonds are as in the CH class. Carbonyl presence means that there is a carbon–oxygen double bond in the compound. Oxygen linkage means that two carbon atoms are bonded together by an oxygen bridge. Ether and the rest of the categories are defined in the conventional manner. In each case, the weight vector was trained to give a positive answer when the value was greater than the cutoff, and a negative dot product when value was equal to or less than that number. All categories of the CHON class were successfully trained. The predictive ability ranged from 68.8 to 98.9%, with an average of 88.0%. Hence, the CHON class is also amenable to successful structural parameter prediction.

While it is not expected that unambiguous molecular structures can always be determined from an empirical treatment of low-resolution mass spectra, information so derived may be of considerable aid in elucidating a compound's structure. A qualitative test of this application was performed. Ten compounds were selected randomly from each of the training and prediction sets of the CH and CHON compounds. The mass spectrum of each test compound was then classified by the trained weight vectors of the applicable class. The classification results were submitted to one of the authors with no additional data. This person then proceeded, as far as possible, to derive the molecular formula and structure of each test compound. Tables 4 and 5 present the learning machine results for the two cases, along with a summary of the conclusions reached and their accuracy. The column in Table 5 labeled incorrect classifications gives the number of weight vectors for each compound that gave erroneous answers, that is, the number of incorrect predictions out of 43 for CH compounds. (These data were included as supplementary information after the deductions had been completed.)

Table 4, for the CH training set, shows that the computations give information which allows correct deduction of the moledular formula in every case. Furthermore, the correct structure was uniquely determined in five cases (nos. 1, 3, 6, 9, and 10), the correct structure was determined as one of two isomers in two more cases (nos. 2 and 4) and one of three in two other cases (nos. 7 and 8) and, in the final case (no. 5), the information was insufficient to reduce the possible structures to a small number, although some conclusions were reached. An illustration of the reasoning used to arrive at the listed results follows, for example, no. 1 of this set. From the carbon number and hydrogen number weight vectors, 5 carbons and 12 hydrogens are indicated. The carbon/hydrogen ratio of $2n + 2$ agrees, and no unsaturation is indicated in the form of double bonds, triple bonds, vinyl groups, benzene, or other rings; hence it is concluded that the molecular formula is C_5H_{12}. Therefore the compound must be either n-pentane, isopentane, or

TABLE 3 CHON Class

	Cutoff	Training set Negative category	Positive category	Spectra feedback	Prediction set Negative category	Positive category	Per cent prediction
Carbon number	9	141	9	49	86	7	94.6
	8	131	19	120	81	12	87.1
	7	124	26	220	76	17	90.3
	6	111	39	266	71	22	83.9
	5	80	70	516	52	41	80.0
	4	48	102	159	35	58	82.8
Hydrogen number	20	146	4	22	92	1	98.9
	19	143	7	73	91	2	98.9
	18	141	9	93	91	2	98.9
	17	138	12	84	90	3	96.8
	16	137	13	74	90	3	96.8
	15	130	20	159	89	4	89.2
	14	127	23	115	89	4	92.5
	13	116	34	279	80	13	86.0
	12	113	37	345	78	15	80.6
	11	92	58	254	66	27	87.1
	10	82	68	229	63	30	82.8
	9	55	95	237	45	48	80.6
	8	49	101	204	39	54	81.7
	7	33	117	80	27	66	81.7
	6	20	130	63	20	73	83.9
	5	14	136	62	10	83	87.1
Oxygen number	2	145	5	46	85	8	94.6
	1	102	48	963	55	38	76.3
	0	43	107	88	26	67	93.5
Nitrogen number	1	146	4	55	87	6	93.5
	0	99	51	67	63	30	91.4
Carbon:hydrogen ratio	$2n + 3$	138	12	26	4	89	97.8
	$2n + 2$	100	50	213	28	65	87.1
	$2n + 1$	87	63	223	32	61	81.7
	$2n$	37	113	112	63	30	91.4
	$2n - 1$	31	119	113	65	28	91.4
	$2n - 2$	22	128	45	71	22	94.6

38

Feature	Level						%
	2n − 3	3	17	133	62	75	18	92.5
	2n − 4	4	15	135	36	79	14	93.5
	2n − 5	5	9	141	19	80	13	93.5
Largest clump	9	146	4	20	90	3	96.8
	8	137	13	96	87	6	88.2
	7	132	18	164	81	12	89.2
	6	125	25	168	78	15	85.0
	5	108	42	123	66	27	86.0
	4	89	61	133	56	37	88.2
	3	53	97	193	37	56	85.0
	2	29	121	75	22	71	90.3
	1	10	140	56	4	89	95.7
Number of clumps	2	136	14	187	78	15	90.3
	1	87	63	544	55	38	75.3
Methyl	2	119	31	279	80	13	80.6
	1	58	92	935	43	50	78.5
	0	23	127	340	11	82	85.0
Ethyl	1	124	26	326	80	13	85.0
	0	90	60	852	52	41	68.8
Largest ring	5	128	22	112	74	19	93.6
	4	114	36	116	66	27	93.6
Number of —C=C—	2	140	10	23	79	14	94.6
	1	132	18	29	73	20	94.6
	0	123	27	145	66	27	88.2
Carbonyl	0	106	44	852	61	32	73.1
Oxygen linkage	0	92	58	582	59	34	75.3
Ether	0	111	39	1005	75	18	76.3
Heteroatom in ring	0	116	34	288	74	19	85.0
Amines	0	112	38	83	73	20	92.5
Alcohol	0	129	21	122	81	12	89.2
Aromatic	0	138	12	16	76	17	94.6
Odd hydrogen number	0	104	46	194	69	24	86.0

TABLE 4 CH Training Set

	Molecular Formula	Carbon/Hydrogen Ratio	Terminal Methyl	Terminal Ethyl	Terminal n-Propyl	Largest Ring	Branch-point Carbons	No. of —C=C—	Carbon without Hydrogen	φ	—C≡C—	Vinyl
1	C_5H_{12}	$2n+2$	3	1	0	None	1	1	0	No	No	No
2	C_6H_{12}	$2n$	3	1	0	None	1	1	1	No	No	No
3	C_5H_8	$2n-2$	2	0	0	None	0	2	1	No	No	No
4	C_7H_{12}	$2n-2$	3	1	0	None	1	0	>1	No	Yes	No
5	$C_{10}H_8$	$<2n-6$	0	0	0	>6	2	>2	>1	No	No	No
6	C_8H_{18}	$2n+2$	4	1	1	None	1	0	1	No	No	No
7	C_8H_{16}	$2n$	2	0	0	6	2	0	0	No	No	No
8	C_8H_8	$<2n-6$	0	0	0	6	2	>2	2	Yes	No	No
9	C_9H_{18}	$2n$	2	0	0	5	2	0	0	No	No	No
10	C_9H_{10}	$<2n-6$	0	0	0	6	1	>2	0	Yes	No	Yes

	Predicted Molecular Formula	Actual Molecular Formula	Predicted Molecular Structure	Actual Molecular Structure
1	C_6H_{12}	C_5H_{12}	CH₃—CH—CH₂—CH₃ (CH)	CH₃—CH—CH₂—CH₃ (CH)

| 2 | C_6H_{12} | C_6H_{12} | $CH_3-CH=\overset{\displaystyle CH_3}{\underset{\displaystyle |}{C}}-CH_2-CH_3$ or $CH_3-\overset{\displaystyle |}{\underset{\displaystyle CH_3}{C}}=CH-CH_2-CH_3$ |

| 3 | C_5H_8 | C_5H_8 | $CH_3-CH=\overset{\displaystyle |}{\underset{\displaystyle CH_3}{CH}}-CH_3$ |

| 4 | C_7H_{12} | C_7H_{12} | $CH_3-CH-C\equiv C-CH_2-CH_3$ or $CH_3-C\equiv C-\overset{\displaystyle CH_3}{\underset{\displaystyle |}{CH}}-CH_2-CH_3$ |

| 5 | $C_{10}H_8$ | $C_{10}H_8$ | Two rings, five double bonds |

| 6 | C_8H_{18} | C_8H_{18} | $CH_3-CH_2-CH_2-CH_2-\overset{\displaystyle CH_3}{\underset{\displaystyle CH_3}{C}}-CH_3$ |

| 7 | C_8H_{16} | C_8H_{16} | or isomers |

| 8 | C_8H_8 | C_8H_8 | or isomers |

| 9 | C_9H_{18} | C_9H_{18} | $CH_2-\overset{\displaystyle CH_3}{\underset{\displaystyle CH_3}{CH}}$ |

| 10 | C_9H_{10} | C_9H_{10} | $CH_2-CH=CH_2$ |

41

TABLE 5 CH Prediction Set

	Carbon number	Hydrogen number	Carbon: hydrogen ratio	Methyl	Ethyl	n-Propyl	Largest ring	Branch point carbons	No. of —C=C—	Carbon without hydrogen	Benzene ring	Triple bond	Vinyl group	Incor. classifications
1	5	10	$2n$	2	0	0	0	2	1	0	no	no	no	3
2	6	14	$2n$	3	0	0	0	1	1	0	no	no	yes	2
3	6	12	$2n$	3	>1	0	0	2	1	>1	no	no	yes	6
4	5	8	$2n-2$	2	0	0	5	1	0,2	0	no	no	no	4
5	10	20	$2n$	4	1	0	6	2	0	1	no	no	no	3
6	8	16	$2n-6, 2n-2$	2	>1	1	5	2	>2	>1	yes	no	no	6
7	8	16	$2n$	1	1	0	6	2	0	0	no	no	no	2
8	8	14	$2n-2$	1	1	0	5	1	0	1	no	no	no	6
9	9	20	$2n+2$	4	1	0	0	2	0	1	no	no	no	2
10	9	18	$2n$	2	0	1	6	1	0	0	no	no	no	2

	Predicted molecular formula	Actual molecular formula	Predicted molecular structure	Actual molecular structure
1	C_5H_{10}		$CH_3CH\!-\!CH\!-\!CH_3,$ with CH above and CH_3 below	$CH_3\!-\!C\!=\!CH\!-\!CH_3$ with CH_3 above
2	C_6H_{12}		$CH_3\!-\!CH\!-\!CH_2\!-\!CH\!=\!CH_2$ with CH_3 above	$CH_3\!-\!CH\!-\!CH_2\!-\!CH\!=\!CH_2$

42

3 C_6H_{12} $CH_3-CH_2-\underset{\underset{CH_2-CH_3}{|}}{C}=CH_2$ or $CH_3-CH_2-CH_2-CH=CH-CH_2-CH_3$

4 C_5H_8 $CH_3-\underset{\underset{CH_3}{|}}{C}=C=CH_2$ or $CH_3-CH=C=CH-CH_3$

5 $C_{10}H_{20}$ 5-membered ring, 3 Me, 1 ethyl

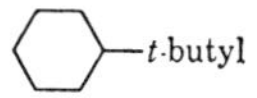

6 $C_{10}H_{14}$ ϕ, xylene

7 C_8H_{16}

cyclopentane with CH_3 and $CH_2-CH_2-CH_3$ or C_2H_5 or isomers

8 C_8H_{14}

cyclohexene with CH_2-CH_3

9 C_9H_{20}

$CH_3-CH_2-CH-CH_2-CH_2-CH-CH-CH_3$ with CH_3 branches

10 C_9H_{18}

cyclohexane with $-CH_2-CH_2-CH_3$ or cyclohexane with CH_3 and C_2H_5

— right column structures —

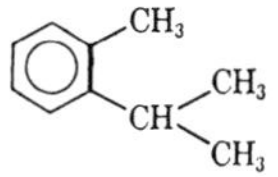

$CH_3-CH=CH-CH$ with CH_3 branches

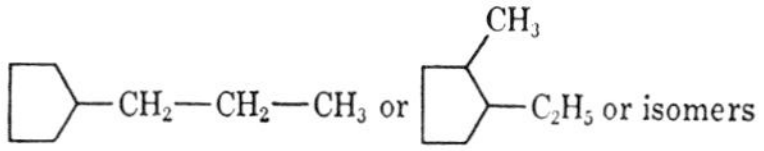

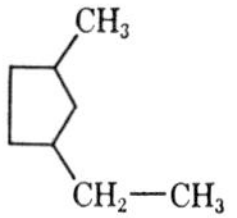

$CH_2=CH-CH-CH=CH-CH_3$

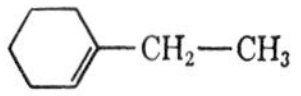

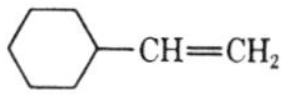

t-butyl cyclohexane

benzene ring with CH_3 and $CH(CH_3)_2$

cyclopentane with CH_3 and CH_2-CH_3

cyclohexane with $CH=CH_2$

$CH_3-CH-CH_2-CH-CH_2-CH_2-CH_3$ with CH_2-CH_3

cyclohexane with $CH(CH_3)_2$

43

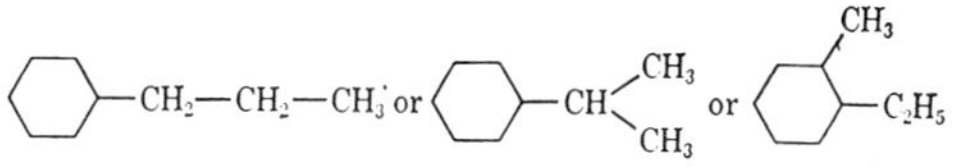

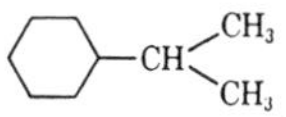

neopentane. The three-methyl conclusion eliminates both n-pentane and neopentane. Confirmation is given by the one ethyl result and the one branch-point carbon result. Finally, the last of an n-propyl also eliminates n-pentane. Hence it is concluded that the compound is isopentane. Note that considerable redundancy exists in the available information. This is particularly useful in the case of prediction set compounds for which erroneous conclusions are much more likely.

For the CH prediction set, as summarized in Table 5, the molecular formula was correctly predicted in all but one case. Less success occurred with determining the structure than with the training set; however, the results were still of considerable interest. One unique structure (no. 2) was predicted correctly, another (no. 7) was correctly predicted as one of two possibilities, and a third (no. 10) was correctly predicted as one of three possibilities. (Isomers based on ring-branching positions were not distinguishable from any of the structural parameters considered, and were therefore not considered separately.) Of the remaining seven examples, five (nos. 1, 3, 4, 8, and 9) were judged to be usefully close to the correct structure in that often the location of a double bond or branch point was the only difference between the predicted and the correct structure. In one case (no. 6) the molecular formula was incorrect, hence there was no chance of a correct structure prediction. However, in this case the results were still useful, because the main structure (a substituted benzene ring) was correctly predicted even though the substituted groups were incorrect. In the final case (no. 5) the computational results were so contradictory as to make very little sense. Hence no structure prediction was attempted in this case. It should be noted that even contradictory data of this sort is useful in that it constitutes a warning not to attempt to use it for prediction.

In the original investigation studies similar to those presented here for the CH class were made for the CHON class of spectra.

TLUs with Nonzero Threshold (3)

The elementary TLUs used above compare the scalar s to zero in order to make classifications. A slight generalization involves using a nonzero threshold Z. Then, when a dot product is formed between the pattern vector $\mathbf{X}$ and the weight vector $\mathbf{W}$, the scalar formed is compared to Z. If s is greater than Z, then the pattern is said to belong to one category; if s is less than $-Z$, then it is said to belong to the other category; and if s is between $-Z$ and Z, then the pattern is not classified. The region between $-Z$ and Z is known as the dead zone. This classification method can be thought of in terms of two hyperplanes, where patterns are classed in one category if they fall on one side of both planes, are classed in the other category if they fall on the other side of both planes, and are not classed if they fall between the planes.

Training is performed using the threshold much as if $Z = 0$. Feedback is applied only to correct errors, but errors include both incorrect classification and failure to make a classification. The new weight vector is calculated from the old one during training with the equation

$$\mathbf{W}' = \mathbf{W} \pm c\mathbf{X} \tag{1}$$

in which $\mathbf{W}'$ is the new weight vector, $\mathbf{W}$ is the old one, $\mathbf{X}$ is the pattern vector being classified, c is the correction increment, and the sign is chosen according to the sense of the error. The correction increment is calculated according to the equation

$$c = \frac{2}{\mathbf{X} \cdot \mathbf{X}} (\pm Z - s) \tag{2}$$

where $\mathbf{X} \cdot \mathbf{X}$ is related to the length of the pattern vector $\mathbf{X}$, s is the scalar result that led to the incorrect classification ($s = \mathbf{X} \cdot \mathbf{W}$), and the sign is chosen according to the sense of the error. This feedback method shifts the decision surface, so that after feedback the pattern point is as far on the correct side of the decision surface as it was on the incorrect side before feedback. With an assiduous choice of Z one would expect predictive ability to rise as compared to the zero threshold case, because one is essentially more narrowly defining the hypervolume within which the hyperplane can fall and still be a "good" decision surface. One also expects an increase in reliability with a nonzero threshold.

Table 6 shows the results of training binary pattern classifiers with and without the threshold. The data used are the same as previously described with reference to Table 1 in Chapter III. For each test, 300 spectra were randomly selected from the data set; three different randomizations were used, resulting in three different training and prediction set populations, as shown at the bottom of Table 6. The TLUs with $Z = 50$ require more feedbacks to train to 100% recognition in each case, and they display higher predictive ability in each case. These TLUs with $Z = 50$ are able to classify correctly 95.0, 95.8, and 96.0% of the complete unknowns comprising the prediction set. (The remainder of Table 6 is described in a few pages.)

The figures given for percent recognition were obtained as follows. After training to complete recognition, the members of the training set were again fed to the pattern classifier one at a time, but they were varied by a randomly generated vector before classification. The magnitude of each component of the randomly generated vector was drawn from a gaussian distribution. The figure refers to the relative standard deviation of the gaussian distribution used. For example, $\sigma = 5\%$ means that approximately one-third of

TABLE 6 Comparison of Properties of Four Pattern Classification Methods. Oxygen Presence–Absence Determination

	Randomiza-tion	Feedback +1WV/−1WV	Percent Prediction +1WV/−1WV	Average Percent Prediction	Percent Recognition	
					σ = 2%	σ = 5%
Lunear machine	1	135/119	96.3/94.0	95.3	98.5	96.4
	2	135/127	95.3/94.0	94.7		
	3	92/172	93.3/94.0	93.6		
Linear machine with threshold Z = 50	1	239/184	96.0/96.0	96.0	99.6	98.6
	2	286/211	96.3/95.7	95.8		
	3	157/179	95.7/94.3	95.0		
Committee machine	1	180		96.0	95.0	90.7
	2	153		94.0		
	3	95		91.7		
Committee machine with threshold Z = 50	1	375		98.0	100.0	99.8
	2	297		96.3		
	3	398		95.7		

	Randomiza-tion	Training Set +/−	Prediction Set +/−
	1	84/216	89/211
	2	92/208	81/219
	3	78/222	95/205

the components in each pattern vector were varied more than ±5% from their original values before variation. The variations were imposed after the logarithmic transformation process, so the occurrence of errors was much more frequent than it would be in an actual laboratory situation in which the errors would be introduced before any transformation. It is seen in Table 6 that the percentage recognition rises when the pattern classifier is trained with the nonzero threshold.

Table 7 shows the results of investigating the properties of binary pattern classifiers as a function of threshold size. The number of feedbacks required to classify all the patterns in the training set correctly rises as a function of Z, as would be expected. The predictive ability and reliability both rise also. The last column gives the number of ambiguous m/e positions found out of the 132 used in the training. As Z increases and the volume in which the final decision surface can fall is more narrowly bounded, and number of m/e positions used by the pattern classifier rises.

When training is done with a threshold, then the threshold can be employed in the prediction process. Table 8 shows the results of such an investigation. Column 2 shows the results obtained by using the trained weight vectors to classify the unknown members of the training set with a threshold of zero. Then the members of the prediction set were classified using the indicated values for Z, that is, not classifying those patterns for which the scalar obtained fell between $-Z$ and Z. The percent prediction figures given are the percentage of classifications attempted that were correct. It is seen that in every case the percent prediction with threshold is higher than that obtained without the threshold. The two sets of prediction figures were obtained with two weight vectors trained with different initializations. While the absolute change in predictive abilities is small in this case, the rise in predictive abilities with Z is a general observation.

Another study of the use of nonzero thresholds yielded a somewhat different formulation of the problem (4). During training a corrected weight vector is calculated from the old one by the equation

$$\mathbf{W'} = \mathbf{W} + c\mathbf{X} \tag{3}$$

The correction increment c is calculated with the previously defined equation:

$$c = \frac{2(\pm Z - s)}{\mathbf{X} \cdot \mathbf{X}} \tag{4}$$

A further restriction can be placed on the performance of the feedback; the weight vector can be required to always have a length of 1, that is,

TABLE 7 Properties of Binary Pattern Classifiers as a Function of Threshold Z

Z	Feedback	Average Feedback	Percent Prediction	Average Percent Prediction	Percent Recognition $\sigma = 2\%$	Percent Recognition $\sigma = 5\%$	m/e Positions Discarded
0	135/118	126	96.3/94.0	95.3	98.5	96.4	31
25	156/150	153	95.3/96.0	95.7	99.3	97.5	32
50	239/184	206	96.0/96.0	96.0	99.6	98.6	27
75	275/203	239	96.0/96.3	96.1	99.7	98.9	22

TABLE 8 Enhanced Predictive Ability of Binary Pattern Classifiers by Using Threshold Prediction Process

Z	Initialization A Average Percent Prediction	Initialization A Percent Prediction with Threshold	Initialization A Not Attempted	Initialization B Average Percent Prediction	Initialization B Percent Prediction with Threshold	Initialization B Not Attempted
25	95.7	96.3	3	96.3	96.3	4
50	96.0	97.0	5	96.7	96.3	7
75	96.1	97.6	8	96.9	96.3	9

$|\mathbf{W}| = 1$. In this case the correction increment must be calculated from the quadratic equation

$$\alpha^2 + \beta c + \gamma = 0 \tag{5}$$

in which

$$\alpha = \mathbf{X} \cdot \mathbf{X}(4Z^2 - 4Zs + s^2 - \mathbf{X} \cdot \mathbf{X})$$

$$\beta = 2s(s^2 - 4Zs + 4Z^2 - \mathbf{X} \cdot \mathbf{X}) \tag{6}$$

$$\gamma = 4Z^2 - 4Zs$$

The solution of a quadratic equation results in two solutions, only one of which is meaningful in this case. It should be noted that, in the event that Z equals zero, the equations reduce to the simpler form.

A training routine was implemented using these training equations with the set of 630 low-resolution mass spectra previously described. The results obtained are described in the following paragraphs.

The addition of positive Z should increase the predictive ability of $\mathbf{W}$ for linearly separable cases by forcing a more optimum decision surface. The results of training with a positive Z on some questions known to be linearly separable are shown in Table 9. Prediction after maximizing Z is compared with predictive ability obtained with $Z = 0$. In each case a training set of 300 patterns was randomly chosen; the remaining 330 spectra were considered a prediction set. Patterns in the prediction set were classified solely according to the sign of the dot product, so that any differences in prediction were attributable to different orientations of the decision surface. The posi-

TABLE 9   Comparison of Prediction Results for $Z = 0$ to Those for a Positive Z

Positive Category	Z_{maxr}	Prediction (% correct) Negative Category	Positive Category	Total
Oxygen and/or nitrogen presence	0.0	77.2	91.8	86.4
	6.0	83.0	93.8	89.7
Oxygen presence	0.0	93.3	80.7	89.7
	4.7	95.6	87.2	93.9
C/H ratio of 1:2	0.0	82.1	77.9	80.6
	0.27	82.1	79.7	81.2
>6 Carbon atoms	0.0	95.7	90.0	92.4
	1.3	95.1	91.5	93.0

tive categories are listed in the table, while the negative categories consisted of all patterns not satisfying the indicated criterion. In all cases Z was forced close to a maximum; however, it was found that, if any improvement occurred in prediction, it was almost wholly realized by 95% of Z_{max}, so that continued iteration was unnecessary. That is, trying to force Z beyond 95% of the maximum possible value is extremely difficult, and gives little or no additional improvement in predictive ability. Table 10 shows the change in prediction with Z for the cases that gave significant improvement. It is noted in Table 10 that overall prediction improves in all cases tried, although not as much in some cases as in others.

A further advantage of Z lies in the significance of the magnitude of the dot product, that is, it is proportional to the distance of the unknown pattern from the decision plane. This provides a measure of confidence in the classification of a pattern according to this distance. Table 11 records illustrative results of number correct as a function of the magnitude of Z_{max}. Although there is considerable noise due to the small numbers, the trend is apparent. Confidence is enhanced greatly for patterns that lie further than Z from the decision surface and becomes correspondingly worse as the magnitude of the dot product decreases. Note the difference in prediction inside the decision width from that outside; also the predictive values for the latter are to be compared to that obtained with the TLU (see Table 9).

The third benefit of the decision surface width results on application of a negative Z to linearly inseparable categories. This is important, because it allows the same algorithms to be used for nonseparable cases as for separable cases.

To illustrate the utility with the mass spectra, categories that were known to be linearly separable when all mass peaks greater than 0.5% were used were made inseparable by retention of only the six largest peaks in each

TABLE 10   Prediction as a Function of Z

Oxygen and/or Nitrogen Presence		Oxygen Presence	
Z	Prediction	Z	Prediction
0.0	86.4	0.0	89.7
4.0	89.1	2.0	91.5
5.0	90.0	3.0	93.6
6.0	89.7	4.0	93.6
6.125	89.1	4.5	93.9
6.25	89.1	4.7	93.9

TABLE 11   Confidence; Number Correct during Prediction as a Function of Distance from the Decision Surface

$\|\mathbf{W} \cdot \mathbf{X}\|$	Oxygen Presence		Oxygen and/or Nitrogen Presence	
	No. Correct/Total	Correct (%)	No. Correct/Total	Correct (%)
$> \not{Z}$	278/285	98	265/276	96
0.8 to 1.0	14/17	82	14/17	82
0.6 to 0.8	4/6	67	6/10	60
0.4 to 0.6	5/6	83	8/11	73
0.2 to 0.4	3/7	43	6/13	55
0.0 to 0.2	6/9	67	2/3	67
Total $< \not{Z}$	32/45	71	36/54	67
Overall	310/330	94	301/330	90.0

spectrum. That is, the pattern set consisted of 630 spectra, each containing only the six most intense peaks in its mass spectrum.

Surprisingly, most of the questions tried using only the six largest peaks were linearly separable and the two examples listed in Table 12 were the only ones in which the use of a negative $\not{Z}$ was required. However, these adequately prove the superiority of the $\not{Z}$ approach when compared with the TLU. In each case prediction is considerably better with the use of a no-decision region. The results for a carbon/hydrogen ratio of 1:2 are particularly striking, especially the difference between negative categories (93.4% with negative $\not{Z}$ as opposed to 72.7% with $\not{Z} = 0$).

TABLE 12   TLU Prediction Compared to That of the Negative Z on Linearly Inseparable Categories

	Prediction (% Correct)			
	C_nH_{2n+2} Question		C_nH_{2n} Question	
	Negative $\not{Z}$	$\not{Z} = 0$	Negative $\not{Z}$	$\not{Z} = 0$
Negative category	91.2	83.2	93.4	72.7
Positive category	72.4	55.3	92.2	84.5
Total	88.0	76.8	92.8	77.1
Number of spectra not classified	55	0	121	0

Committee Machine (3)

All the studies cited above used a single TLU to classify patterns. A slight generalization involves using two levels of TLUs. In this method the pattern to be classified is presented simultaneously to three (or five or seven, etc.) TLUs that operate in the normal manner. However, the outputs of the first layer of TLUs go to a second-layer TLU which follows a majority rule law. Thus the whole committee of classifiers sends its output to the vote taker which classifies the original pattern into the cateogry agreed upon by the majority of the first level TLUs.

A useful and simple training method is to make only the minimum necessary error correction feedback at each step in the training. Thus, when an incorrect classification is made during training, the weights of the TLUs that missed being correct by the smallest amount are altered using the usual feedback equation. Only enough TLUs are changed to ensure correct classification after the feedback.

Table 6 shows the results of training a committee machine with three members with the same three sets of data used by the other pattern classification implementations. Convergence to 100% recognition is seen to be fast, and the predictive ability is high. However, this committee machine is evidently quite sensitive to variations in the training set members, as seen from the percent recognition figures.

A pattern classification system can be developed in which each member has a nonzero threshold. Training is as before—the scalars of each member of the committee must be outside the dead zone to register any classification, and the vote taker registers an overall classification only when a majority of the committee members agrees on a classification. One would expect this to be somewhat more powerful pattern classification system than the simpler systems.

Table 6 shows the results of training a committee of three TLUs with $Z = 50$ and with the same three sets of data as the other implementations in the table. The number of feedbacks is seen to rise, although it remains at a reasonable level. However, the percent prediction is much better. This pattern classifier correctly classified 95.7, 96.3, and 98.0% of the members of the prediction set, which were complete unknowns. The percentage recognition is seen to be very high also; this implementation is nearly impervious to errors in patterns of this size.

TLUs with Electrochemical Data

Another entirely different chemical data set to which the simple TLUs were applied is SEP curves (5, 6). The data set used and the preprocessing applied to the data were described in detail in Chapter III. One-hundred-thirty-three features per pattern were developed. A program was used which performed the following operation. Two weight vectors were trained—one was initialized with all $+1$ and the other with all -1. Features were eliminated whose trained weight vector components had opposite signs for the two weight vectors. The process could be repeated.

A training set of SEPs was generated with the following characteristics. For the one-component patterns, n values of 1.00, 1.02, . . ., 2.98, 3.00 were used together with random peak locations, resulting in $(3)(101) = 303$ possible patterns. For the two-component patterns, n values of 1.0, 1.4, 1.8, 2.2, 2.6, 3.0 were used, peak separations of 8, 10, and 12 mV were used, and peak height rations of 20:1, 10:1, 1:1, 1:10, and 1:20 were used. There were 540 possible two-component patterns, of which 533 were used. The sampling conditions used were that E_p was observed to 0.1 mV, and all potentials were evaluated relative to it at 2-mV intervals. The observed peak height was normalized to 1.0. Then each SEP had its 133 features extracted to make up the training set.

A prediction set of SEPs was generated as follows. For one-component patterns, 100 n values of 1.01, 1.03, . . ., 2.97, 2.99 were used, 200 n values in the range 1.000 to 2.999 were selected randomly, and random peak locations were used. For two-component patterns, n values were randomly selected between 1.000 and 3.000, peak separations were randomly selected between 8.000 and 12.000 mV, 300 peak height ratios were randomly selected between 1:1 and 1:20, and 300 peak height ratios were randomly selected between 20:1 and 1:1. Thus 300 one-component and 600 two-component patterns were generated. They were sampled in the same manner as the training set patterns, and each had the 133 features extracted.

Table 13 shows the training results and predictive ability obtained with these sets of SEPs. Five cycles of the training-feature selection cycle described above are shown. An undecided response occurs during training when the two weight vectors being trained classify a pattern in opposite categories. An error is the result of incorrect classification of a pattern by both weight vectors. Percent accuracy is the percentage of correct classifications relative to the number of classifications made.

Table 13 demonstrates that the two weight vectors developed a substantial ability to decide between one-component and two-component SEPs. Recognition of the members of the training set did not decline with feature selection from 133 to 57 features, remaining at approximately 96% accuracy with 5 to 6% undecided. The results are poorer for prediction, but follow the same trend.

The study summarized in Table 13 was followed by further studies of the

TABLE 13 Recognition of Training Set and Prediction of SEP Patterns

Cycle No.	No. of Features	No. of Iterations	Training				Prediction			
			One-Component		Two-Component		One-Component		Two-Component	
			Undecided (%)	Accuracy (%)	Undecided (%)	Accuracy (%)	Undecided (%)	Accuracy (%)	Undecided (%)	Accuracy (%)
1	133	50	19.5	98.8	10.1	94.1	25.7	97.8	10.0	89.1
		100	13.2	95.4	8.8	96.1	15.3	94.1	8.8	91.6
2	75	50	14.9	97.7	6.8	95.8	16.7	96.0	6.5	90.5
3	66	50	12.5	98.9	8.1	96.3	13.7	96.9	4.3	89.0
4	59	50	5.6	90.9	3.6	96.9	11.0	92.1	2.2	89.9
5	57	5	6.6	89.7	5.1	87.1				
		10	9.6	94.5	5.6	91.8				
		15	8.9	96.7	5.1	92.5	11.3	97.7	5.2	88.6
		20	7.6	95.0	3.9	93.4	8.3	94.6	3.0	88.1
		25	7.3	96.8	3.0	93.6	5.0	96.1	1.5	88.3
		30	7.6	97.5	3.9	94.5	6.7	98.2	2.3	88.9
		35	9.6	97.4	4.9	95.3	9.0	97.8	2.3	89.2
		40	6.3	96.5	4.7	95.9	10.3	97.4	2.0	89.2
		45	5.0	92.0	3.0	96.3	6.7	92.9	2.2	89.8
		50	6.3	93.0	3.0	97.1	10.7	92.2	1.7	90.0
		75	5.0	91.7	2.1	96.9	6.3	87.2	2.0	90.6
		100	3.3	91.5	1.7	96.4	5.3	88.0	1.7	89.8
		125	7.3	93.2	2.4	96.1	8.0	88.4	1.7	90.0
		150	7.3	94.3	2.1	96.0	7.0	88.9	2.3	90.4

original work, and the number of features was decreased further. However, the point to be made here is that TLUs can be used to detect the presence of doublet SEPs under widely varying conditions not necessarily amenable to visual interpretation.

Iterative Least-Squares Training

A study has developed an alternative method for the development of chemically useful weight vectors utilizing an iterative least-squares method (7). The patterns are fit to a nonsingular function with linear parameters. There are many processes by which this can be done. The method employed was the linearization or Taylor series method (8). It uses the results of linear least squares in a succession of stages.

The nonlinear function used in this work is the hyperbolic tangent. It was chosen because it is well suited for pattern dichotomizers. The properties of this function compare favorably with the ordinary least-squares multicategory method:

$$R = \sum_{i=1}^{N} (s_i - Y_i)^2 \tag{7}$$

in which Y_i is the correct numerical answer, s_i is the scalar product, i denotes the ith pattern, and N is the number of patterns used in the training set. The ordinary least-squares multicategory method has also been applied to chemical data interpretation (9).

Y_i is set equal to $+1$ if the ith pattern belongs to category 1, and -1 if the ith pattern belongs to category 2. Then a weight vector $\mathbf{W}$ must be found such that the discriminant function $g(\mathbf{X}_i)$ will be positive if Y_i has a value of $+1$, and negative if Y_i has a value of -1. The components of $\mathbf{W}$ are to be developed through the linearization least-squares technique.

The function through which the patterns are fitted is $F(s_i) = \tanh s_i$, where s_i is the scalar product for the ith pattern, $s_i = \mathbf{W} \cdot \mathbf{X}_i$. The hyperbolic tangent function was used because it is positive for all positive values of the independent variable and negative for all negative values of the independent variable. Thus the overall procedure is equivalent to minimizing the number of disagreements in sign between $F(s_i)$ and Y_i. According to the least-squares principle then, the function to be minimized becomes

$$Q = \sum_{i=1}^{N} [Y_i - F(s_i)]^2 \tag{8}$$

Thus the weight vector $\mathbf{W}$ must be found such that Q is a minimum. N is the number of patterns used as the training set.

In order to set up an iterative procedure with $s_i^0 = \mathbf{W}^0\mathbf{X}_i$ as a starting condition, $\tanh s_i$ is expanded in terms of the Taylor series up to and including the first derivative. The expansion contains most of the value of the function. Hence

$$\tanh s_i = \tanh s_i^0 + \sum_{j=1}^{d+1} \frac{\partial \tanh s_i}{\partial w_j}\bigg|^0 dw_j \tag{9}$$

which in terms of the chain rule gives

$$\tanh s_i = \tanh s_i^0 + \sum_{j=1}^{d+1} \operatorname{sech}^2 s_i^0 x_{ij}\, dw_j \tag{10}$$

in which i stands for the ith pattern and j stands for the jth component. Hence the function to be minimized now assumes the form

$$Q = \sum_{i=1}^{N} \left(Y_i - \tanh s_i^0 - \sum_{j=1}^{d+1} \operatorname{sech}^2 s_i^0 x_{ij}\, dw_j \right)^2 \tag{11}$$

The minimum for one iterative step is achieved for all such vectors $d\mathbf{W} = dw_1, \ldots, dw_{d+1}$ that satisfy the minimization principle:

$$\frac{\partial Q}{\partial w_k} = 0 \qquad \text{for } k = 1, \ldots, d+1 \tag{12}$$

Hence

$$0 = -2 \sum_{i=1}^{N} \left(Y_i - \tanh s_i^0 - \sum_{j=1}^{d+1} \operatorname{sech}^2 s_i^0 x_{ij}\, dw_j \right) \operatorname{sech}^2 s_i^0 x_{ik} \tag{13}$$

which can be expressed in terms of linear equations:

$$\sum_{i=1}^{N} (Y_i - \tanh s_i^0)\operatorname{sech}^2 s_i^0 x_{ik} = \sum_{j=1}^{d+1}\sum_{i=1}^{N} \operatorname{sech}^4 s_i^0 x_{ij} x_{ik}\, dw_j \tag{14}$$

In terms of matrices these equations become

$$\mathbf{A}\, d\mathbf{W} = b \tag{15}$$

in which

$$a_{jk} = \sum_{i=1}^{N} \operatorname{sech}^4 s_i^0 x_{ij} x_{ik} \tag{16}$$

and

$$b_k = \sum_{i=1}^{N} (Y_i - \tanh s_i^0)\operatorname{sech}^2 s_i^0 x_{ik} \tag{17}$$

so that the system

$$\mathbf{A}\, d\mathbf{W} = b \tag{18}$$

must be solved for the solution vector $d\mathbf{W}$.

The matrix $\mathbf{A}$ is real symmetric, and positive definite. Hence it is non-singular and has a unique nontrivial solution. This solution can be obtained by any suitable method, either direct or iterative, for solving a set of linear equations.

Since $F(s_i)$ is dependent on the value of W and W is the desired solution, then the solution vector $\mathbf{W}^{(1)}$ starting from $\mathbf{W}^{(0)}$ cannot be accurate. The linearization method requires setting $\mathbf{W}^{(1)} = \mathbf{W}^{(0)} + d\mathbf{W}^{(1)}$, and then beginning a second iteration. The process is repeated until $\mathbf{W}^{(l+1)} = \mathbf{W}^{(l)} + d\mathbf{W}^{(l+1)}$ satisfies the condition

$$|d\mathbf{W}^{(l+1)}|/|\mathbf{W}^{(l+1)}| < \epsilon \tag{19}$$

That is, the ratio of the norm of $d\mathbf{W}^{(l+1)}$ to the norm of $\mathbf{W}^{(l+1)}$ is less than some arbitrarily chosen small quantity ϵ. When this condition is met, the system has become self-consistent, no benefit is gained by further computation, and the method terminates.

In this study the domain of F was limited to the closed interval $(-2.5, +2.5)$. The range of the function $F(s_i)$ for this interval is $(-0.987, +0.987)$. The initial values of the weight vector were chosen in such a way as to ensure that $|s_i| < 2.5$. When this condition is met, the rate of convergence is speeded up, since the change occurring in the components will not be large and therefore $\mathbf{W}$ will not be subject to large errors.

Many direct methods for solving sets of linear equations are available. The method used in this study consisted of Gauss-Jordan elimination with complete pivoting. The subroutine evaluated the inverse of the matrix, the determinant, and the solution vector.

The data employed in this study consisted of 450 low-resolution mass spectra from the API Research Project 44 tables as obtained from the United Kingdom Atomic Energy Authority on magnetic tape. Each spectrum consisted of a number of intensities listed in sequential order according to the m/e position. The lowest intensity reported was 0.01% of the largest peak in the spectrum. There are 132 m/e positions that have at least 10 peaks in them out of the 450 spectra; therefore 132 m/e positions is the upper limit on d, the dimensionality of the pattern vectors. The spectra were all of small organic molecules with molecular formulas $C_{1-10}H_{1-22}O_{0-4}H_{0-2}$.

The data were transformed into a normalized form. Previous experience showed that, if the ratio between the highest and lowest intensity of each spectrum was lowered from its existing value of $100.0/0.01 = 10,000$, the system converged much more readily. Hence each intensity was normalized according to the equation

$$I' = 10 \log_{10} I1000 \tag{20}$$

in which I is the original intensity and I' is the normalized intensity. (I' corresponds to x_{ij}, where i denotes the spectrum number and j denotes the m/e index.) For storage purposes the values of I' were rounded off to the nearest integer. With the intensities in normalized form the new range of intensities become 10 to 50, and therefore the ratio of the highest to the lowest intensity was 5.

In this study two sets of data were formed from·the overall data set—the training set and the prediction set. The training set usually consisted of 150 random selected spectra which were used by the learning machine to develop a suitable set of weight vector components. The prediction set consisted of some or all of the remaining data and was used to test how well the learning machine could classify patterns it had never previously encountered.

The iterative least-squares method was applied to the determination of oxygen in small organic molecules. The problem had been feature-selected and the number of features had been reduced from 132 to 31 with complete recognition and a prediction percentage of 93.9%.

For this problem Y_i was set to $+1$ if the ith spectrum contained oxygen, and to -1 if the ith spectrum did not contain oxygen. The weight vector components were initialized so that, if w_j had value p, w_{j+1} was assigned value $-p$. In this problem the value of w_1 was set equal to $+0.01$ or to -0.01. These initial values were found to be adequate in maintaining the values of the scalar products within the desired interval $(-2.5, +2.5)$. Minimization of the distance between the clusters has been found to facilitate convergence.

Table 14 presents the results obtained for the oxygen presence determination problem both with the original 31 peaks and after further feature selection. Recognition is nearly complete, and prediction percentage is of the order of 98%, regardless of the number of features employed.

TABLE 14 Oxygen Presence Problem

m/e Position	Training Set Population +	Training Set Population −	Percent Recognition	Prediction Set Population +	Prediction Set Population −	Percent Prediction
31	42	108	99.3	26	74	98
	39	111	100.0	81	219	96
	43	107	98.7	77	223	98
22	43	107	99.3	77	223	98
18	43	107	99.3	77	223	98

TABLE 15 Nitrogen Presence Problem

m/e Position	Training Set Population +	Training Set Population −	Percent Recognition	Prediction Set Population +	Prediction Set Population −	Percent Prediction
31	12	138	99.3	20	280	95
	8	142	100.0	24	276	97
	9	141	99.7	23	277	94.7
	11	139	99.3	21	279	96.7
21	8	142	98.7	24	276	96
	17	133	98.7	15	285	97
	20	130	98.0	12	288	97
14	12	138	98.7	20	280	97

The second test problem was the determination of the presence or absence of nitrogen atoms in small organic molecules. The number of features had been reduced from 132 to 43. Of these 43 features, those corresponding to the lowest 31 m/e positions were chosen. As for oxygen, Y_i was set equal to $+1$ if the ith spectrum contained nitrogen, and to -1 if it did not contain nitrogen. The initial weight vector components were chosen as before.

Table 15 shows the results obtained for the nitrogen problem with the original 31 peaks and after further feature selection. Again there is no loss in prediction or recognition ability. A prediction percentage of 96.7% was obtained for 31 features, and of 98% for 21 features.

Piecewise Linear TLUs

A different approach to using several TLUs to make a binary decision is demonstrated in reference 10. In this work a series of linear TLUs was used together in a piecewise linear array.

The mathematical formulation of this procedure is described in the following discussion.

A number of categories R relating to the information sought in the data, for example, molecular structure parameters, is chosen, and it is assumed that members of a given category are arranged together in the pattern space. The actual distribution for each of the R categories may be quite complex and may exhibit several local maxima corresponding to the subcategories L_i, $i = 1, \ldots, R$. A simple two-dimensional example (excluding the $d + 1$ dimension) is depicted in Figure 1, with $R = 2$, $L_1 = 3$, and $L_2 = 1$.

The classification problem is then to develop a set of discriminant func-

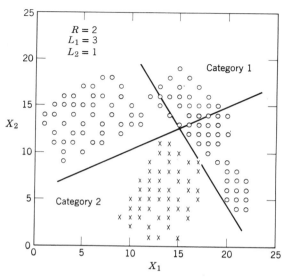

Figure 1. *Two-dimensional classification problem with* $R = 2$, $L_1 = 3$, *and* $L_2 = 1$.

tions s_i (in this case piecewise linear functions) describing d-dimensional hyperplanes which segregate the members of a given category from those of all other categories. The most extensively studied piecewise linear discriminant functions have the form

$$s_i = \max_{j=1,\ldots,L_i} s_i^{(j)} \qquad \text{for } i = 1, \ldots, R \qquad (21)$$

in which the $s_i^{(j)}$ are known as subsidiary discriminant functions defined by

$$s_i^{(j)} = f_i^{(j)}(\mathbf{X}) = w_{i1}^{(j)}x_1 + w_{i2}^{(j)}x_2 + \cdots + w_{id}^{(j)}x_d + w_{i,\,d+1}^{(j)}x_{d+1} \qquad (22)$$

When such an arrangement is used, a pattern is assigned to category k if s_k has the largest value of all the discriminant functions s_i, $i = 1, \ldots, R$.

An iterative training process using an error correction method adjusts the individual weights comprising the weight vectors $\mathbf{W}_i^{(j)}$ until all patterns are correctly classified. Suppose that a pattern belonging to the kth category is presented to the classifier and that s_l has the largest value among the discriminant functions. The feedback procedure is:

$$\mathbf{W}_k' = \mathbf{W}_k + c\mathbf{X}$$
$$\mathbf{W}_l' = \mathbf{W}_l - c\mathbf{X} \qquad (23)$$

in which c is the positive correction increment. Several rules for choosing c

are in use. A variant of the fractional correction rule was chosen for this work. Under the fractional correction rule, the increment c is always chosen such that the decision surface defined by the weight vectors $\mathbf{W}_k$ and $\mathbf{W}_l$ is moved a fixed fraction λ of its normal distance to the pattern point $\mathbf{X}$; thus, for $\lambda = 2$, the new decision surface defined by $\mathbf{W}_k'$ and $\mathbf{W}_l'$ lies an equal distance on the opposite side of the pattern point $\mathbf{X}$. In cases in which R is greater than 2, a complication arises due to the varying magnitudes of the weight vectors; since the lengths of the weight vectors enter into the evaluation of $s_i^{(j)}$, the decision process is biased toward the longer weight vectors. Such a bias is avoided if the weight vectors are normalized to unit length; normalization adds the restrictions.

$$\mathbf{W}_k' = \frac{\mathbf{W}_k + c_k\mathbf{X}}{|\mathbf{W}_k + c_k\mathbf{X}|} \quad \text{and} \quad \mathbf{W}_l' = \frac{\mathbf{W}_l - c_l\mathbf{X}}{|\mathbf{W}_l - c_l\mathbf{X}|} \tag{24}$$

Substitution of these equations into those above leads to quadratic equations in c_k and c_l:

$$c_k^2[(\mathbf{XX})^2 - \mathbf{XX}s_l^2] + c_k[2s_k(\mathbf{XX} - s_l^2)] + s_k^2 - s_l^2 = 0$$
$$c_l^2[(\mathbf{XX})^2 - \mathbf{XX}s_k^2] + c_l[2s_l(\mathbf{XX} - s_k)] + s_l^2 - s_k^2 = 0 \tag{25}$$

The values of c_k and c_l must be obtained from these equations for each feedback correction.

While the number of categories R is usually known when the classification problem is formulated, the number of subcategories L_i is, in general, unknown. The conventional method for dealing with this uncertainty is to assign an adequate but fixed number of weight vectors to each category in advance of the actual training process. Although such a procedure leads to a solution, it allows for the possibility of introducing unnecessary calculation both in the training process and in classification tasks after training, if the number of vectors exceeds the number actually required; in addition, the use of an execssive number of weight vectors can produce "overfitting" of the data distribution of the training set, which leads subsequently to poor performance on unknown patterns.

In designing the present classifier, the intention was to provide a simple means for internal generation of new weight vectors as dictated by the training process itself; thus, hopefully, the pattern classifier would evolve to that level of complexity just sufficient to provide the desired solution. Any pattern classifier of a given complexity, which uses an error correction training procedure, exhibits oscillatory behavior when presented with an insoluble problem; a classifier using a single linear discriminant function, for example, could never converge to a solution for the problem illustrated in Figure 1; instead, the orientation of the single decision surface would undergo major oscillations for an indefinite period. It is logical then to look for means of detecting such oscillations which indicate the need for a more complex decision surface.

The method proposed here involves periodic evaluation of a function based on the quantity

$$\sum_{m=1}^{M} \frac{s_k - s_l}{M} \tag{26}$$

Here M is the total number of patterns in the training set; it is understood that k represents the category of the pattern m, and that s_l is the largest discriminant function among the s_i, $i \neq k$. Thus, if the pattern m is correctly classified, it will contribute a positive term to the sum; if the pattern is mis-classified, that term will be negative. Although no rigorous argument can be offered for the use of the above quantity, particularly as to how its abso-lute magnitude should vary during training, the relative' changes in its value were found empirically to reflect oscillations in the decision surface and were deemed worthy of investigation.

The actual function evaluated has the form:

$$P_t = \frac{1}{\tau} \left(P_{t-1} + \sum_{m=1}^{M} \frac{s_k - s_l}{M} \right) \tag{27}$$

The function P is therefore a "running weighted integral" of the past values of the summation, with a time constant τ. A function of this general form has a smoothing effect, because of a decreasing contribution of earlier incre-ments I with time as governed by the time constant.

A simple criterion for addition of another subsidiary discriminant func-tion is:

$$P_t \geq \alpha P_{t-1} \qquad \text{no addition}$$

$$P_t < \alpha P_{t-1} \qquad \text{addition}$$

where α is a positive fraction, $0 < \alpha \leq 1$. The parameters τ and α determine the sensitivity of the function P to changes in decision surface orientation and are adjusted experimentally.

Figure 2 illustrates the behavior of P during training on the two-dimen-sional problem of Figure 1. The pattern classifier begins with a single hyperplane decision surface, but since a single plane is insufficient, the value of P eventually falls off at the interval $t = 4$; at this point, $P_4 < P_3$, and another subsidiary discriminant function is introduced, which allows the training process to converge to a solution. The dotted lines in Figure 2 indicate the result of suppressing the addition of a new subsidiary function.

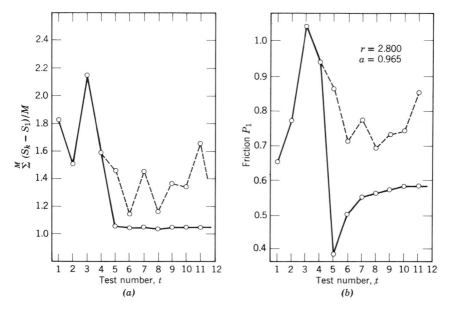

Figure 2. (a) Blot of $\Sigma^M (S_k - S_1)/M$ versus test number for classification problem of Figure 1. (b) Plot of function P versus test number for classification problem of Figure 1.

The data used in this work were taken from the API Research Project 44 tables and consisted of 387 low-resolution mass spectra of hydrocarbons of formulas $C_{1-10}H_{4-22}$. Each spectrum was a series of paired numbers giving the nominal mass of the fragment and the intensity of the peak. Peaks with intensities of less than 1 percent were discarded. The average number of peaks per spectrum was approximately 35.

Generally, sets of 200 spectra were randomly chosen for training, and the remaining 187 spectra were used to test the predictive ability of the trained classifier.

An interesting test of the piecewise classifier and its ability to evolve in complexity is presented by the problem of detecting the presence or absence of a carbon–carbon double bond, an important structural feature. A double bond may occur in one of several interesting ways in a given molecule. A simple binary classifier fails to converge in less than several thousand feedbacks during training on this problem, and prediction with the resulting weight vector is approximately 75%. The results of training with the piecewise linear classifier are shown in Table 16 which lists the values of τ and α, the number of classifications during training, the number of feedbacks, the number of weight vectors arrived at and, finally, the predictive ability of the trained classifier. The data listed for each trial in Table 16 represent averages for five separate runs, each using different weight initializations and order of pattern presentation. Convergence was obtained in all runs, and the number of classifications may be taken as roughly proportional to length of training.

In each trial the classifier begins with only two weight vectors, and subsequently adds vectors as necessary; the effect of varying the time constant τ on the final number of weight vectors is evident; larger values of τ increase the sensitivity of P to oscillations and therefore yield more weight vectors. (The seemingly large number of vectors in trial 5 results from pairwise addition of weight vectors, one from each category.) The results show predictive abilities ranging from roughly 78 to 87%, and improvement of 3 to 12% over the binary classifier.

In the original investigation further improvement of the capabilities of the program was made in several ways, but the basic self-evolving machine was shown to be quite good.

MULTICATEGORY PATTERN CLASSIFIERS

The work discussed to this point has all dealt with binary classifications. Now we turn to the development of multicategory classification ability. This ability is important in that many chemical questions have multicategory answers, for example, the number of functional groups per molecule. One means of achieving multicategory classification is through the use of an array of binary pattern classifiers. Several types of arrays can be used, and these are discussed in the following sections.

Branching Tree of Binary Pattern Classifiers

Figure 3 shows a branching tree of binary classifiers to determine molecular formulas, which are described in reference 11. Twenty-six weight vectors were developed to dichotomize formula subscripts correctly for the mass spectra of 346 compounds of formulas $C_{1-7}H_{1-16}O_{0-3}N_{0-2}$. By storing each of these weight vectors (each requiring the same dimensionality as the original patterns), the molecular formula could be computed using the binary decision tree shown in Figure 3. This constitutes a storage saving of better than a factor of 10 over retaining all 346 spectra. It is not meant to imply, however, that the entire information content of the mass spectra has been compressed into the weight vectors, but rather the necessary set of information to determine molecular formulas; that is, by specifying the question to be answered, it was possible to reduce the amount of data to be retained.

TABLE 16 Performance of Self-Generating Piecewise Classifier on Carbon–Carbon Double Bond Problem

| | | | | | | | Prediction | | |
| | | | | | | | | Average | |
Trial	Tau	Alpha	Classifica-tions	Feedbacks	Weight Vectors	Range	Overall	Category 1	Category 2
1	2.80	0.985	6925	930	3–4	78.6–86.1	81.9	82.3	81.5
2	3.10	0.985	6680	892	4	78.1–85.0	81.0	82.5	79.6
3	3.50	0.985	7623	1085	3–5	77.5–86.6	81.4	83.1	79.9
4	4.50	0.985	6159	914	5	79.1–86.6	81.7	84.8	79.2
5	2.00	0.990	8898	12.6	6–10	78.6–82.9	81.5	84.1	78.9

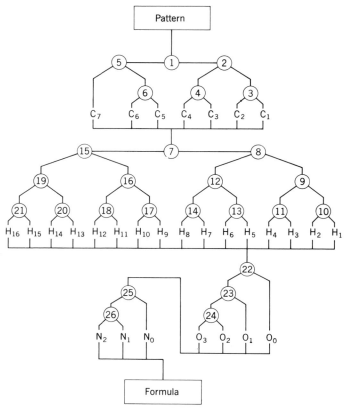

Figure 3. Branching tree of binary classifiers.

Furthermore, the molecular formula could then be "computed" by a series of dot product calculations rather than from a library search. (Such a computation takes about 50 ms with a second-generation computer or 30 min with a desk calculator.)

Another more recent study of multicategory classifications also used low-resolution mass spectra (12). Low-resolution mass spectra were obtained from a collection purchased on magnetic tape from the Mass Spectrometry Data Centre, Atomic Weapons Research Establishment, United Kingdom Atomic Energy Commission. Six-hundred spectra, pertaining to compounds of molecular formulas $C_{3-10}H_{2-22}O_{0-4}N_{0-2}$, were taken from that portion of the tape comprised of API Research Project 44 spectra. The spectra were randomly divided into a training set of 200 and a prediction set of 400. A second data set, consisting of the hydrocarbon subset of the 600 spectra, was also used. Five C_3 hydrocarbons were deleted because of the low population of this category. The remaining 372 C_4 to C_{10} hydrocarbons were randomly divided into a training set of 200 and a prediction set of 172. The same training and prediction sets were used throughout the investigation, so that the results of the various classification schemes could be compared.

On-hundred-thirty-two m/e positions were of significance, considering the entire set of spectra, and therefore each spectrum was represented as a 132-dimensional pattern. The original peak intensities, normalized with respect to the highest peak in each spectrum, fell within the range 0.01 to 99.99. The intensities were renormalized based on the total ion current, or sum of intensities for each spectrum, in order to place all spectra on the same intensity scale. Subsequent logarithmic transformation gave intensities of 10 to 59.

In the hydrocarbon set 3.5% of the spectra exhibited no parent peak (intensity $<0.001\%$ of total ion current). An additional 8.6% of the spectra had parent peaks of 0.001 to 0.1% intensity, and another 12.1% of intensity 0.1 to 0.5%. With a substantial portion of the spectra lacking a parent peak discernible from noise, prediction of carbon number for hydrocarbons is not a trivial problem.

The binary pattern classifiers are arranged in a branching network. The method was applied to the hydrocarbon data set, each binary classifier being trained to dichotomize a set of pattern vectors according to the scheme presented in Fig. 4. In developing the six binary pattern classifiers, only those spectra pertinent to each branch point were utilized. For example, the weight vector for branch point 3 was trained using only those spectra in the overall training set of 200 that correspond to carbon number 4 or 5. The predictive ability of each weight vector was tested on a similarly chosen set of spectra from the overall prediction set of 172. The results of such training

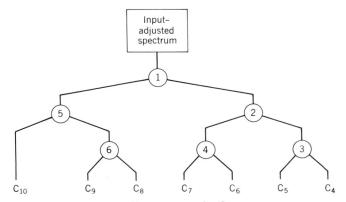

Figure 4.   Branching tree of binary pattern classifiers.

TABLE 17 *Binary Pattern Classifiers for Branching-Tree Array. Hydrocarbon Set*

Branch number[a]	Training set			Prediction set		
	Negative category	Positive category	Number of feedbacks	Negative category	Positive category	Per cent prediction
1	81	119	83	67	105	98.3
2	18	63	23	24	43	100.0
3	7	11	3	6	18	100.0
4	29	34	25	17	26	100.0
5	80	39	67	74	31	97.1
6	42	38	31	35	39	97.3

[a] Branch numbers defined in Figure 1.

and prediction are given in Table 17. The weight vectors were then used in a master, branching-tree program; a predictive ability of 95.4% resulted.

The branching-tree classification was also carried out using weight vectors trained by considering the entire training set for each branch point, that is, the weight vectors for the parallel scheme above. Only in the case of the first branch point (cutoff of seven) are the two types of weight vectors identical. The resulting overall prediction was 94.2% for this second method.

The branching-tree arrangement of parallel weight vectors was also applied to the entire data set of 600 spectra to give 76.3% prediction. Here a symmetrical tree of seven binary pattern classifiers was necessary in order to classify into eight categories, C_3 to C_{10}.

Parallel Arrangement of Binary Pattern Classifiers

A series of binary classifiers can also be used as a group of ever-increasing thresholds. For example, in the previous figure separations to distinguish between 4 and 10 carbons are accomplished by putting carbons 4 through 7 in one class and 8 through 10 in the other. Next those classes were broken down as shown in the branching tree. However, another method is to define six decision makers as follows. The first TLU has carbon 4 in one class by itself and carbons 5 through 10 in the other class, the second TLU has carbons 4 and 5 in one class and 6 through 10 in the other, and so on. The parallel form is less efficient than the branching-tree arrangement in that it requires more computation to perform a classification. However, there is greater redundancy in the classification procedure, and this usually results in better overall predictive ability. This latter procedure also has some ability to point out unknowns that are poorly represented by the training set. This might be suspected when the classifiers radically disagree with one another on the proper classification of an unknown.

Figure 5 illustrates the application of a parallel array of binary pattern classifiers to the determination of carbon number using the same data set as in the previous example. In this method each binary pattern classifier is trained to classify pattern vectors into one of two categories separated by a cutoff point. The positive category is comprised of those pattern vectors having carbon numbers greater than the cutoff, and the negative category contains those with a carbon number less than or equal to the cutoff. The classifier with a cutoff of 7, for example, was trained to give a positive dot product for carbon numbers 8, 9, and 10, and a negative dot product for carbon numbers 7 and less.

The binary classifiers, each with a different cutoff, were trained and their individual predictive abilities determined. The results are summarized in Table 18. Here, and throughout this investigation, weight vectors for each

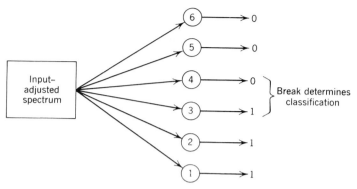

Figure 5. Parallel arrangement of binary pattern classifiers.

question were trained in two ways: starting with initial weights w_j's of all
$+1$, and of all -1. The trained weight vector giving the better predictive
ability was saved.

The set of trained weight vectors was used in a master program to predict
carbon number as illustrated in Figure 5. The sequence of n binary decisions
may be written as an n-digit binary number containing 0 for negative classifications and 1 for positive classifications. The carbon number is then indicated by the cutoff value giving the first negative classification. For example,
in the C_4 to C_{10} case the number 000000 corresponds to carbon number four;
000111, to carbon number seven. More than one discontinuity, for example,
000101, indicates that one or more erroneous decisions were made and the
pattern cannot be validly classified. Ninety-three percent of the unknown
spectra in the prediction set were correctly classified for the hydrocarbon
data set, and 76.0% for the entire data set.

Binary Code Classification

In order to classify pattern vectors into a number of categories m, the
parallel and branching-tree schemes require $m - 1$ binary classifiers. An
alternate method is to use n classifiers, where $2^{n-1} < m \leq 2^n$, such that when
each decision (0 or 1) is used as a digit in a binary number the decimal
equivalent is the proper category number. Up to eight categories can be
accommodated by a three-digit binary number $d_3d_2d_1$. For example, 000 $\equiv$
category 1 (C_3), 001 $\equiv$ category 2 (C_4), . . . , 111 $\equiv$ category 7 (C_{10}). For
the hydrocarbon set, where there are only seven categories, two such codes
are possible— the above one, designated variation 1, and another one called
variation 2, where 000 $\equiv$ C_4, . . . , 110 $\equiv$ C_{10}. Three binary classifiers were

TABLE 18 *Binary Pattern Classifiers for Parallel Array*

Cutoff	Training set			Prediction set		
	Negative category	Positive category	Number of feedbacks	Negative category	Positive category	Per cent prediction
Hydro-carbons						
9	161	39	93	141	31	97.7
8	123	77	61	102	70	98.3
7	81	119	83	67	105	98.3
6	47	153	51	41	131	99.4
5	18	182	24	24	148	100.0
4	7	193	3	6	166	99.4
Entire data set						
9	179	21	91	335	65	92.8
8	151	49	180	272	128	97.0
7	117	83	115	217	183	97.3
6	85	115	124	171	229	95.5
5	59	141	149	101	299	95.5
4	29	171	101	52	348	95.0
3	8	192	71	21	379	98.3

TABLE 19 *Binary Pattern Classifiers for Binary Codes*

Bits	Carbon numbers in positive category	Training set			Prediction set		
		Negative category	Positive category	Number of feedbacks	Negative category	Positive category	Per cent prediction
Hydrocarbons							
Variation 1 ($C_4 \equiv 001$)							
d_1	10 8 6 4	83	117	395	83	89	87.8
d_2	10 9 6 5	83	117	157	67	105	95.9
d_3	10 9 8 7	47	153	51	41	131	99.4
c_1	9 8 5 4	102	98	450	74	98	92.4
c_2	9 7 6 4	92	108	535	84	88	91.3
c_3	8 7 6 5	84	116	79	76	96	98.3
c_4	10 7 5 4	109	91	603	91	81	93.6
Variation 2 ($C_4 \equiv 000$)							
d_1	9 7 5	117	83	395	89	83	87.8
d_2	10 7 6	98	102	450	98	74	92.4
d_3	10 9 8	81	119	83	67	105	98.3
c_1	10 9 6 5	83	117	157	67	105	95.9
c_2	10 8 7 5	74	126	649	62	110	89.5
c_3	9 8 7 6	57	143	82	55	117	95.9
c_4	8 6 5	118	82	323	102	70	93.6
Entire data set							
d_1	10 8 6 4	98	102	2158	179	221	72.8
d_2	10 9 6 5	95	105	1369	153	247	80.0
d_3	10 9 8 7	85	115	124	171	229	95.5
c_1	9 8 5 4	87	113	909	202	198	76.8
c_2	9 7 6 4	93	107	2088	190	210	73.8
c_3	8 7 6 5	78	122	404	180	220	90.3
c_4	10 7 5 4	96	104	1460	209	191	78.3

trained to classify appropriate sets of carbon numbers for each variation as given in Table 19. The results of training and testing the individual predictive abilities are also listed. Overall prediction of carbon number in the hydrocarbon set was 83.7% and 80.2% for variations 1 and 2, respectively, and 57.8% for the entire data set.

The accuracy of the binary number can be improved by introducing additional digits to form an error-correcting Hamming code (13, 14). If the evaluated binary number $d_n \ldots d_1$ differs from the correct number by an error in only one digit, that is, if the Hamming distance is 1, the error can be detected and corrected by k check bits, such that $2^k \geq n + k + 1$ for n data bits. The Hamming distance is thus increased from 1 in the original n-digit number to 3 or more in the $(n + k)$-digit number. In the present application three check bits are needed for the three data bits—a Hamming (6, 3) code. Each check bit c_i is constructed so that, when all data bits are correct, the parity of the check bit and two associated data bits, considered collectively, is even. When this is the case, the parity bit p_i is given a value of 0. For odd parity $p_i = 1$. The three sets of associated bits are:

$$p_1: c_1 + d_1 + d_2$$
$$p_2: c_2 + d_1 + d_3$$
$$p_3: c_3 + d_2 + d_3$$

An error in the data bits gives rise to odd parity in two of the parity groups. The binary number given by the parity digits $p_3 p_2 p_1$ then assumes a nonzero value, and the decimal equivalent is the position of the erroneous bit in the composite number $d_3 d_2 p_3 d_1 p_2 p_1$. For example, in terms of variation 1, the six-digit number for carbon number 8 is 100100. If d_1 erroneously equals 0, the two parity checks involving d_1 (p_1 and p_2) become odd, giving the number 100011. $p_3 p_2 p_1 = 011$ has a decimal equivalent of 3, indicating that the third digit is in error and consequently should be a 1 rather than a 0.

The additional binary classifiers needed for error correction c_1, c_2, and c_3, were trained using appropriate subsets of the training set for the positive and negative categories. These sets and the results of training and prediction are given in Table 19. Carbon number prediction by the (6, 3) codes gave increased prediction of 93.0 and 93.6% for variations 1 and 2, respectively, using the hydrocarbon set, and 68.5% using the entire data set.

Since the check bits have less than 100% prediction, it is advantageous to apply error correction only when the data bits are indeed erroneous. Otherwise, errors in the parity bits may lead to misclassification. An additional parity check bit c_4 may be used to detect errors in the data bits by testing their overall parity. Thus, if the sum (modulo two) $d_3 + d_2 + d_1 + c_4$ is even, $p_4 = 0$; if odd, $p_4 = 1$. Even parity indicates that zero or two bits are in error; odd parity indicates that one or three bits are in error. The assumption is made that three errors are unlikely. Then, error correction should be implemented only when the overall parity is odd ($p_4 = 1$). Otherwise, the carbon number is simply predicted from $d_3d_2d_1$. The overall parity check bits were trained and tested as summarized in Table 19. The resulting Hamming (7, 4), codes gave carbon number predictions of 95.4% for variation 1 and 92.4% for variation 2 hydrocarbon data set, and 67.0% with the entire data set.

Carbon number prediction by the various classification schemes is summarized in Table 20. It should be noted that random guessing of the carbon number would be successful one time in seven (14.3%) for the C_4 to C_{10} data set, and one time in eight (12.5%) for the C_3 to C_{10} data set. All percent predictions reported are much greater than the success rate for random guessing, indicating that a substantial amount of learning has taken place. The hydrocarbon results are better in all cases than for the entire data set, which is reasonable considering the more diverse nature of the latter. For a given data set, the percent prediction is similar for all methods of classification with the exception of the three-bit binary codes. Nevertheless, meaningful comparisons can be made.

In the parallel classification scheme, the vast majority of misclassifications were the result of an error in only one of the six or seven binary pattern

TABLE 20   Summary of Carbon Number Prediction by Different Classification Schemes

Classification scheme	Per cent prediction	
	Hydro-carbons[a]	Entire data set[b]
Parallel	93.0	76.0
Branching tree	95.4	. . .
Branching tree (parallel w_j's)	94.2	76.3
Binary codes		
Variation 1: 3-bit	83.7	57.8
6-bit	93.0	68.5
7-bit	95.4	67.0
Variation 2: 3-bit	80.2	
6-bit	93.6	
7-bit	92.4	

[a] C_4-C_{10}; random guessing would give a success rate of 1/7 or 14.3%.
[b] C_3-C_{10}; random guessing would give a success rate of 1/8 or 12.5%.

classifiers. Indeed, all misclassifications were of this type for the hydrocarbon data set, where the number of incorrect classifications was equal to the number of erroneous binary decisions. Ten of the errors were on the boundary between 0 and 1 in the binary word, while in two cases there was an imbedded 0 (no decision made). With the entire data set, 10 of the 96 misclassifications resulted from more than one erroneous binary decision. Imbedded 0's occurred three times. The results are consistent with the general principle that the number of misclassifications by an array of binary pattern classifiers can be no more than the sum of the binary errors and no less than the number of errors made by the best binary classifier.

Use of the weight vectors trained for parallel decisions in the branching-tree scheme led to improved prediction for both data sets. The two spectra in the hydrocarbon data set that had imbedded zeroes in the string of parallel decisions were correctly categorized. This result arose from a fortuitous choice of branch points. Had the alternate mode of branching been used (branch point 1 spearating carbon numbers 10, 9, 8, and 7 from 6, 5, and 4), the percent prediction would have been identical for the parallel and branching-tree arrangements. With the entire data set, one of the three spectra with imbedded zeroes in the binary word was correctly classified. Thus, since the branching tree utilizes only a portion of the binary pattern classifiers for each classification, the resulting predictive ability is always greater than or equal to that for the parallel system with the same weight vectors and prediction set.

Prediction by the branching-tree method was better using the weight vectors trained to dichotomize unique subsets of the training set than using the parallel weight vectors trained on the total training set. This is not surprising, since each branch point encounters a set of pattern vectors that is much more akin to the training sets in the former than in the latter.

All the weight vectors needed for classification by the different binary codes were found to train completely within a reasonable number of feedbacks, that is, the various sets of carbon number were linearly separable—a result that had been far from obvious beforehand. Carbon number prediction by three-bit binary codes proved to be the least accurate method for both data sets. Most misclassifications arose from only one error in the three data bits, suggesting that high overall prediction through error correction is a possibility. All error-correcting codes did in fact lead to increased prediction. It is interesting that, for variations 1 and 2, the predictive ability of the three data bits differed by 2.5%, but the error-correcting portions $c_1c_2c_3$ had comparable accuracy and therefore the percent predictions after error correction differed by only 0.6%.

Faulty check bits in the six-bit codes led to less than optimum prediction, both from failure to correct errors properly and from correct data bits being rendered incorrect. The overall parity check bit of the Hamming (7, 4) code eliminated some of these incorrect "corrections," but here again less than perfect predictive ability led to difficulty. With the hydrocarbon data set, c_4 had the same predictive ability for both carbon number variations but influenced the overall prediction in different ways. For variation 1, the percent prediction was improved slightly, while for variation 2 the predictive ability dropped to the lowest value obtained for any of the methods employing six or seven binary classifiers. For the entire data set, the seven-bit code gave slightly poorer prediction than the six-bit code. Hence the success of the single error-correcting, double error-detecting code is highly dependent on which members of the prediction set give an erroneous decision for the overall parity check.

The highest predictive ability was attained with the hydrocarbon data set using the branching tree and one version of the Hamming (7, 4) code. With the entire data set, the branching-tree method showed the best prediction. For other classification problems and data sets, it appears that one of these two methods should prove optimal. However, in the last analysis the empirical approach of testing the methods and comparing the results is necessary.

Implementation of the branching-tree and error-correcting codes involves three steps: sorting the members of the data set into appropriate subsets, training a binary pattern classifier to classify each selected training set, and overall prediction by means of the combined binary pattern classifiers. (It should be noted that any binary pattern classifier can be used in these arrays. Increased predictive ability could be achieved for the individual binary pattern classifiers, for example, by employing nonzero thresholds during training, or by using layered threshold logic units.) The parallel method is easier to carry out because the sorting step is not necessary.

In training binary pattern classifiers, it is advantageous to have the members of the training set evenly distributed between positive and negative categories, so that classification is not biased toward one of the two choices. The parallel method suffers in this respect. As can be seen in Table 18, except for intermediate cutoffs, the distributions are inherently unbalanced, particularly for the first and last cutoffs. The positive and negative categories are more evenly populated in the branching-tree method (Table 19), because some of the members are excluded in the selection process. The binary method, however, uses all members of the training set for each classifier, and the positive/negative distributions are well balanced (see Table 20). Furthermore, as the number of overall categories, for example, number of carbon atoms, is increased, the distributions approach the optimum value of 50%, even if the populations of individual carbon numbers differ markedly.

The error-correcting binary code, parallel, and branching-tree methods each required six binary classifiers to classify the data set in eight categories. As prediction is extended to a larger number of categories, such as the question of hydrogen number or carbon number in a less restricted data set, binary codes would be easier to implement and might give the best prediction, since the number of required binary classifiers would be minimized. The necessary number of binary pattern classifiers for the branching-tree and parallel methods increases linearly with the number of categories, whereas the number of classifiers increases much less rapidly for the binary codes. Addition of one more binary classifier (d_4) to the Hamming codes of this investigation, for example, would accommodate 15 categories of carbon number. Additional studies on molecular weight prediction show that a Hamming $(10, 4)$ or $(11, 5)$ code can successfully classify the C_4 to C_{10} hydrocarbon set in 47 categories. The branching-tree and parallel arrays require 46 binary classifiers, or over four times as many. The $(10, 4)$ code gave the higher overall prediction of 80.8%, remarkably better than the success rate for random guessing of 1 out of 47, or 2.1%.

The potential usefulness of binary codes in solving chemically significant problems depends on whether the pertinent binary pattern classifiers can be trained to give sufficiently high prediction.

Piecewise Linear Classification

Piecewise linear classifiers can be used to make multicategory classifications. In reference 10 experiments of this type were done. The same data set of low-resolution mass spectra described above in the previous section on piecewise linear classification was used for these experiments as well.

Table 21 shows the results of training a piecewise linear classifier on carbon/hydrogen ratio $(2n + 2, 2n, 2n - 2, 2n - 4, 2n - 6)$; the results were compared with a binary classifier. The overall predictive ability of the multicategory classifier averaged 97.2% for five trials; when broken down into individual categories, the prediction ranged from 98.3% for $2n + 2$ compounds to 92.4% for $2n - 4$ compounds. By comparison, the predictive performance obtained by independently training five binary classifiers for the specific carbon/hydrogen ratios was poorer, particularly for those categories containing only a small number of patterns. For example, while a binary vector trained for $2n - 4$ compounds versus all other yielded an overall prediction of 96.1%, the percentage of patterns actually having the $2n - 4$ ratio that were correctly classified was only 65.3%. Thus, given the spectrum of an unknown $(2n - 4)$ compound, the multicategory classifier

TABLE 21   *Multicategory Application: Carbon/Hydrogen Ratio; Comparison by Category with Binary Classifier*

Multicategory				Binary				
Overall	By category		Prediction set composition	Binary question	Overall	By category		Prediction set composition
						Pos.	Neg.	Pos. Neg.
97.2	$2n+2$	98.3	45/187	$2n+2$ vs. other	97.4	95.4	98.0	44 143
(average	$2n$	97.1	73/187	$2n$ vs. other	94.6	96.4	94.1	74 113
of 5	$2n-2$	98.2	34/187	$2n-2$ vs. other	96.2	92.8	97.1	37 150
trials)	$2n-4$	92.4	10/187	$2n-4$ vs. other	96.1	65.3	96.9	7 180
	$\leq 2n-6$	95.9	25/187	$\leq 2n-6$ vs. other	98.0	88.6	98.8	16 171

provides a decision nearly 30% more reliable than the binary classifier. It is likely that a bias toward the larger category arises in all binary cases in which the numbers of patterns for the two categories are extremely unequal.

K-Nearest-Neighbor Classification

A second pattern recognition technique which is intrinsically a multicategory method is the K-nearest neighbor (Knn) method. Reference 15 describes the KNN method as applied to the interpretation of NMR spectra.

In the KNN method a pattern is classified according to the majority rule of its K nearest neighbors in n space. If the pattern being classified is an unknown, then only training set patterns are used as nearest neighbors. The computation is normally done by calculating and scanning a distance matrix. The distance can be defined as the euclidean distance in n space between points i and j:

$$d_{ij} = \left[\sum_{k=1}^{n} (x_{ik} - x_{jk})^2 \right]^{1/2} \tag{28}$$

Alternatively, any other distance criterion desirable can be used, for example, a weighted distance.

A further application of the KNN technique is given in reference 16. In this work gas chromatographic liquid phases were classified based on Kovats' index data from several different test solutes. The distance between the ith and the jth gas chromatographic liquid phases was calculated from the equation

$$d_{ij} = \left[\sum_{k=1}^{m} (\Delta I_{ik} - \Delta I_{jk})^2 \right]^{1/2} \tag{29}$$

ΔI is the difference between the Kovats index of the compound run on squalane and the phase of interest. For each liquid phase the most similar liquid phase was said to have the smallest d value, and the least similar had the largest d value. The distance measure was used as a guide to determine whether or not one particular phase could be replaced by another. A table of the distances between 226 common liquid phases and the nearest of 12 preferred phases and the relevant distances was given.

Least-Squares Classification

A third method for making multicategory classifications is the least-squares procedures (9).

The weight vector (W) developed in a binary pattern classifier is actually a linear combination of some or all of the patterns of the training set developed in such a fashion that the dot product of the weight vector and ith pattern ($W \cdot X_i$) gives a scalar s_i whose sign indicates to which of two categories the pattern belongs. The principle of the multiclassification method is to develop a weight vector that produces values of s_i whose magnitudes, as well as signs, place the patterns in one of several categories. The correct category s_i^* for each pattern of the training set is defined by some arbitrary value. For example, in the case of developing a weight set to differentiate between compounds having zero, one, two, three, or four oxygens, the values of 0, 1, 2, 3, or 4 may be assigned to the s_i^*'s for the corresponding compound. A least-squares procedure is employed to select the set of weights that computes values of s_i so as to minimize $(s_i - s_i^*)^2$:

$$Q = \sum_{i=1}^{n} (s_i - s_i^*)^2 = \sum_{i=1}^{n} \left(\sum_{j=1}^{m} w_j y_{ij} - s_i^* \right)^2 \qquad (30)$$

where m = number of mass positions, n = number of patterns in the training set, and Q = sum of squares of deviations.

The normal equations to minimize Q are developed by taking the partial derivatives of Q with respect to each weight and setting these equal to zero:

$$\frac{\partial Q}{\partial w_k} = 2 \sum_{i=1}^{n} \sum_{j=1}^{m} (w_j y_{ij} - s_i^*) (y_{ik}) = 0 \qquad (31)$$

for $k = 1, 2, 3, \ldots m$.

The normal equations are solved in the conventional fashion to determine the optimum w_j's and hence W by the least-squares criterion.

For cases in which the number of categories is small compared to the number of patterns, as in the determination of oxygen number in mass spectra, each s_i is considered in the category of the nearest meaningful value. For example, if oxygen numbers zero, one, and two are given s_i^* values of

0, 1, and 2, respectively, an s_i of 1.68 is classified as meaning two oxygens. However, arbitrarily selecting 1.5 as the discriminating value between one and two oxygens may not give the best results. Hence, after the lease-squares step, a "best line" routine is used to compute the discrimination line between the two categories that gives the best answer for the training set. In cases in which large numbers of categories exist, such as for molecular weight determination, the category is no longer as important as the nearness of s_i to the correct answer. Also, the best line calculation may be quite lengthy in cases containing many categories, such as hydrogen number, and has little value for these.

In actual time the least-squares calculation increases as a square of the dimensionality of the problem and at least as the first power of the number of patterns. For a small number of categories it is usually much slower than a series of binary dichotomizers; however, because the calculation length is independent of the number of categories, the least-squares method becomes much more practical as the number of categories increases. The least-squares method, however, is applicable only when the categories of classification are quantizable.

All data were taken from the API Research Project 44 tables, and only intensities equal to or greater than 1% of the maximum peak recorded. The implementation of the least-squares pattern classifier was developed in two stages with different-sized data sets, because of the capabilities of the two computer systems used. The first study dealt with methods of maximizing recognition, or the ability to classify correctly previously seen data, with some emphasis on minimizing the nceessary computation. The data set included 130 low-resolution mass spectra for $C_{1-5}H_{1-12}O_{0-2}N_{0-2}$, with 79 mass positions which along with the $d + 1$ term gave an 80-dimensional system.

The example used for the study was determination of oxygen number as zero, one, or two. Best line adjustments were made after every calculation. Except where indicated, the peak intensities were adjusted to make the maximum peak 100, and then each value was raised to the one-half power.

A least-squares calculation as described above produced a weight vector which correctly classified 123 of the 130 spectra for a 94.6% success in the oxygen zero, one, or two case. (The best line correction improved the calculation from nine to seven wrong.) Of course, in this three-category case random guessing yields a predictive ability of 33%.

The least-squares technique has also been applied to the determination of hydrocarbon types and the structure of the average molecule is a complex mixture such as gasoline (16). A set of equations was set up, one equation per compound in the file of library spectra:

$$y_i = a_0 + \sum_{j=1}^{n} a_j x_{ij} \tag{32}$$

in which i goes from 1 to m, m is the number of compounds in the reference spectra file, that is, the training set size x_{ij} is the peak intensity of the $m/e = j$ peak for compound i, y_i is the value of the property of interest for compound i, and a_j is the coefficient of the weight vector to be used for determining a particular property of the mixture.

If the training set size m is greater than $n + 1$, the number of adjustable parameters, then the parameters can be determined by least squares. The authors point out that the set of equations is general and the x_{ij} can be from diverse sources if desirable. However, such data as is put into the x_{ij}'s must be additive for a mixture. The y_i's to be used can be any property of interest in the mixture to be dtermined. Examples include the percentage of a single compound present, the percentage of a group of compounds such as aromatics, or the number of methyl groups present. For each property, that is, for each set of y_i values, a new set of a_j values must be calculated. Then the a_j values can be put into the equation above to obtain the value of each property for a mixture, given the set of measurements x_j for the mixture.

This least-squares approach was applied to the determination of some features of the average molecule in gasoline. The data set used included mass spectra, NMR spectra, and refractive index and density data, with emphasis on mass spectra. A total of 342 hydrocarbons in the range C_5 to C_{12} was used to develop the system, which was then tested with synthetic mixtures. The results were reported and discussed extensively. It was pointed out that this method yielded carbon and hydrogen percentages by weight that agreed extremely well with the results of standard combustion values. It was also noted that this approach is best suited to analytical determinations in which the number of components in the mixture exceeds the number of analytical measurements made on the mixture, and that nearly singular matrices have to be inverted.

REFERENCES

1. P. C. Jurs, B. R. Kowalski, T. L. Isenhour, and C. N. Reilley, *Anal. Chem.*, **41,** 690 (1969).
2. P. C. Jurs, B. R. Kowalski, T. L. Isenhour, and C. N. Reilley, *Anal. Chem.*, **42,** 1387 (1970).
3. P. C. Jurs, *Anal. Chem.*, **43,** 22 (1971).
4. L. E. Wangen, N. M. Frew, and T. L. Isenhour, *Anal. Chem.*, **43,** 845 (1971).
5. L. B. Sybrandt and S. P. Perone, *Anal. Chem.*, **43,** 382 (1971).
6. P. C. Jurs, *Jap. Anal. (Bunseki Kagaku)*, **21,** 1276 (1972).

7. Lucio Pietrantonio and P. C. Jurs, *Pattern Recognition*, **4,** 391 (1972).

8. R. O. Duda and P. E. Hart, *Pattern Classification and Scene Analysis*, Wiley, New York, in press.

9. B. R. Kowalski, P. C. Jurs, T. L. Isenhour, and C. N. Reilley, *Anal. Chem.*, **41,** 695 (1969).

10. N. M. Frew, L. E. Wangen, and T. L. Isenhour, *Pattern Recognition*, **3,** 281 (1971).

11. P. C. Jurs, B. R. Kowalski, and T. L. Isenhour, *Anal. Chem.*, **41,** 21 (1969).

12. W. L. Felty and P. C. Jurs, *Anal. Chem.*, **45,** 885 (1973).

13. W. W. Peterson, *Error-Correcting Codes*, MIT Press, Cambridge, Mass., 1972.

14. F. E. Lytle, *Anal. Chem.*, **44,** 1867 (1972).

15. B. R. Kowalski and C. F. Bender, *Anal. Chem.*, **44,** 1405 (1972).

16. J. J. Leary, J. B. Justice, S. Tsuge, S. R. Lowry, and T. L. Isenhour, *J. Chromatogr. Sci.*, **11,** 201 (1973).

17. D. D. Tunnicliff and P. A. Wadsworth, *Anal. Chem.*, **45,** 12 (1973).

CHAPTER V

Feature
Selection

In many pattern recognition problems the patterns from the different classes are so mixed that nonlinear discriminant methods must be used to effect a separation. It is preferable to divide the overall pattern classification problem into two parts, the first of which consists of simplifying the problem so that the second part can be effective. Thus the paramount goal of feature selection is dimensionality reduction with no loss of discrimination. With proper and efficient feature selection, the dimensionality of the data being handled is lowered to the point where the ease of implementing various discrimination functions is increased.

Feature selection can be performed on data in two contexts:

1. To reduce the number of raw data to a manageable number before discrimination is attempted

2. To investigate pattern vectors to determine which descriptors are the most important to the classification task

An example will clarify the distinction. In collecting the mass spectra of a group of organic compounds of molecular weights in the neighborhood of 300, one collects the information from m/e 300 to approximately m/e 12. There would be 288 descriptors. But it is obvious that m/e positions where peaks never occurred are not useful and only would interfere, in that these m/e positions increase the dimensionality of the pattern space but carry no information. Therefore, such m/e positions are dropped from inclusion in the pattern vectors. This is the first type of feature selection mentioned. The second would be pursued if it were desired to try to determine which subset of the m/e positions was correlated with a specific chemical question. Since the focus of this discussion is on binary classification, a typical chemical question is whether a spectrum corresponds to a compound containing a ring or not. Then feature selection procedures can be used to try to discover which m/e positions are important for answering this question and which are not.

The first sort of feature selection uses only information contained in the data, for example, statistical parameters calculated from the data. This processing is closely allied with preprocessing and can properly be considered part of preprocessing. The second sort of feature selection additionally uses the results of discriminant training as criteria for the selection of important features.

The major theoretical difficulty in feature selection, as in preprocessing, is that the results obtained must be evaluated by the output of the classification stage of the pattern recognition system. Thus one must deal with the whole system, and this introduces additional complexity.

It must be pointed out that the only guaranteed method for finding the optimum subset of m features from a set of n is to try all $\binom{n}{m} = n!/(n-m)!m!$ possible contributions. This is precluded for data sets of even moderate size, so heuristic methods are required. Generally, the methods used on chemical data are ad hoc, and have been used because they were found to be useful in pursuing the goals mentioned above.

Many types of feature selection have been described in the pattern recognition literature (1–5). However, only a few types have been applied to chemical data. This is primarily because pattern recognition involving chemical data has been nonparametric, as a result of the nature of the data being investigated. Most feature selection methods are primarily useful for data sets for which the distribution functions are known or can be estimated. For example, while the use of principal components analysis is widespread in the field of pattern recognition in general, it has not been used with chemical data because it is primarily intended to be used with random variables and for cases in which important information is related to the variance.

The following discussion considers work in which feature selection methods were used with a selection of chemical data.

An early attempt at feature selection of low-resolution mass spectrometry data appeared in reference 6. The data and least-squares method used to train the weight vectors have been described in detail in Chapter IV in the section on least-squares classification. The spectra have 80 m/e positions coded.

The example used for the study was the determination of oxygen number as zero, one, or two. A least-squares calculation produced a weight vector which correctly classified 123 of the 130 spectra for a 94.6% success in the oxygen zero, one, or two case. (The best line correction improved the calculation from nine to seven wrong). Feature selection was then investigated.

Although any action that decreases the available data and simultaneously the number of adjustable parameters makes recognition more difficult, the calculation time is decreased as well. Hence, even decreasing the dimensionality by a factor of 2 greatly decreases the computation time. Choosing the mass positions to eliminate is a difficult problem requiring the impractical calculation of every possible combination to guarantee the ideal choices. However, it is reasonable to expect that those that least affect the calculation are logical choices. Hence two methods were investigated: (1) eliminating masses on the basis of smallest weight values, and (2) eliminating masses on the basis of smallest cumulative effect on the decision process. In the second case the product R_j, formed by the weight times the sum of the amplitudes of the corresponding mass position, was used. This was an attempt to calculate quantitatively the contribution each feature of the data made to the overall decision process. Features with lesser contributions were discarded. Figure 1 shows the effect of casting out masses in groups of 15 and recalculating with each smaller set.

The two methods seem comparable, although there is some indication that the R_j criterion is better. In considering the R_j method, it is interesting that elimination of the first 45 mass slots, leaving only 35 possible slots (of which the average spectrum has about 15 peaks), has little effect on recognition ability. Even eliminating all but 4 masses and the $d + 1$ term gives 70.0% right answers, which is still far better than random (33%). A further test eliminating masses in groups of 30 by the R_j criterion gave answers of 71.5% recognition after two iterations (60 eliminated), which compared to the four iterations each eliminating 15 masses shows that good results may be obtained without much care in the selection of eliminated masses.

These same two methods of feature selection using weights and R_j's are discussed in reference 7 in conjunction with error correction feedback binary pattern classifiers. They were also employed in work dealing with diverse spectroscopic data outlined in reference 8, which was discussed in Table 2 of Chapter III.

WEIGHT-SIGN FEATURE SELECTION (9)

The feature selection routines described above were algorithmic, and the number of features to be discarded at each stage were determined in advance. There was no capacity in the algorithms for deciding when to terminate; features were discarded even when the pattern recognition system's performance was thereby substantially degraded. This condition led to investi-

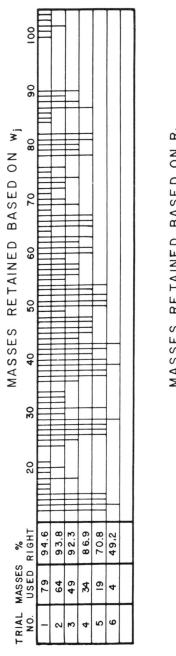

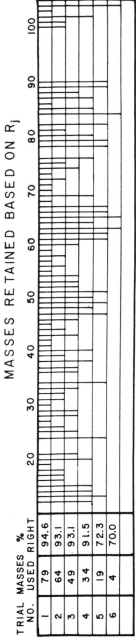

Figure 1. Recognition as a function of number of eliminated masses.

gations which developed a dynamic feature selection method whose outcome depends on conditions during its execution. That is, the method discards features until it can find no more features that are not contributing substantially to the overall decision process and then terminates. The method has been successfully applied to mass spectra, and before describing it, the data pool employed must be described.

The mass spectral data used in the work were drawn from the API Research Project 44 tables. The data consist of 630 low-resolution mass spectra of relatively small organic molecules of the formula $C_{1-10}H_{1-24}O_{0-4}N_{0-2}$. Only peaks with intensities greater than 1% of the largest peak in the spectrum were used; most of the spectra had 15 to 40 such peaks, and a total of 17,137 peaks occurred in all. The intensities were normalized by taking the square root.

Out of the data pool of 630 spectra, 300 were randomly selected as a training set, and the remaining 330 spectra were reserved for use as a prediction set. Within the entire data pool, the highest m/e position for which a spectrum exhibited a peak was $m/e = 195$. Thus $d = 195$ and X and Y had to be 195-dimensional vectors. However, the linear decision surfaces being used treated each dimension independently, with no interaction between terms. Thus the dimensionality was reduced to 155, because no spectrum exhibited peaks for any of the other 40 m/e positions. A further reduction resulted from discarding the 36 m/e positions of which fewer than 10 peaks occurred in the entire data pool. There were 17,137 peaks total, of which only 111, or 0.6%, were discarded by this selective procedure, leaving 119 m/e positions. Tests showed that the positions thus discarded would have made no appreciable contribution to ease of classification if they had been included.

Mass spectra data consisting of 119 features per pattern were investigated by a feature selection program based on a linear learning machine method. The sequence of operations is as follows. Two weight vectors are to be trained to detect oxygen presence or absence. One of the weight vectors is initialized with all components set equal to $+1$, and the other one is initialized with all components set to -1. The two weight vectors are then trained to classify the members of the training set. Their predictive abilities are tested, and then the m/e positions not contributing to the classification are discarded by the following scheme. The components of the two weight vectors that correspond to the same m/e position are compared; if both components have the same sign, the m/e position is said to correlate well with oxygen absence if they are both negative, and oxygen presence if they are both positive, and that m/e position is retained. But m/e positions for which the signs are different are considered ambiguous and are discarded. After all the m/e positions have been checked, the entire cycle begins anew with the spectra of reduced dimensionality. The process is repeated as long as the feature extraction routine can find m/e positions that are ambiguous.

Table 1 shows the results of applying the feature selection procedure to oxygen presence or absence classification. The pattern classifiers are trained to detect oxygen presence in the compound yielding each spectrum regardless of the type of oxygen group present. The populations of the training set and prediction set are given at the bottom of the table. The training set contained 82 spectra of compounds that contain oxygen and 218 that do not. The first column in Table 1 gives the number of m/e positions being considered at each stage of the feature selection process. Column 2 gives the number of feedbacks performed while training to complete recognition of the members of the training set by the weight vector initiated with $+1$. Column 4 is the same parameter for the weight vector initiated with -1. Columns 3 and 5 give the predictive percentages exhibited by the two weight vectors on the 330 members of the prediction set for each stage of the feature selection process.

Through seven iterations, the number of features per pattern was reduced from 119 to 37 m/e positions. Despite this decrease in dimensionality, the number of feedbacks performed during training remained approximately constant; the total computer time used in training thus fell, because each classification involved a shorter calculation in a lower dimensional space. The average predictive ability remained high, even with only 37 out of the original 195 m/e positions being considered.

Table 2 shows the results of another test conducted identically to that of Table 1, except that a somewhat different training procedure was used. The

TABLE 1. Training for Oxygen Presence

m/e positions	+1 weight vector		−1 weight vector		Av % prediction
	Feedbacks	Predic-tion, %	Feedbacks	Predic-tion, %	
119	236	92.1	219	93.3	92.7
69	179	93.6	221	93.3	93.4
51	208	94.9	223	94.9	94.9
43	202	94.2	256	94.9	94.5
40	217	92.7	217	94.2	93.4
38	184	93.9	203	92.7	93.3
37	199	94.6	213	93.9	94.2
31	235	94.6	202	93.3	93.9

Av 93.8

	+	−	Total
Training set	82	218	300
Prediction set	92	238	330

TABLE 2. Training for Oxygen Presence

m/e positions	+1 weight vector		−1 weight vector		Av % prediction
	Feedbacks	Predic-tion, %	Feedbacks	Predic-tion, %	
119	210	90.0	208	93.3	91.7
74	187	93.3	174	92.7	93.0
53	179	93.3	187	92.7	93.0
45	165	92.7	281	92.1	92.4
42	161	92.1	221	93.0	92.5
38	158	92.1	223	93.3·	92.7
31	210	93.0	192	93.9	93.5
					Av 92.7

difference in the training procedure was that the members of the training set were presented in a different sequence, thus yielding different decision surfaces. The results are comparable to those in Table 1. After the two feature selection routines had been allowed to reduce the number of features per pattern to 37 and 38 m/e positions, a list of features common to both lists was compiled. Thirth-one m/e positions were selected by both feature selection routines. Then both routines were allowed to train using these 31 m/e positions. Neither routine could find any more ambiguous m/e positions and they both terminated with the results shown at the bottom of Tables 1 and 2. It is interesting to note that in both cases the ability of the classifiers to categorize correctly complete unknowns was nearly the highest observed for any training sequence, 93.9 and 93.5%. It appears that removal of ambiguous m/e positions from the problem does not degrade the classifier's performance.

Of the 31 m/e positions selected by the feature selection process, 14 have positive weight vector components, that is, they correlate with oxygen presence. Table 3 lists these 14 m/e positions along with the fragments that correspond to each position, taken from a published compilation of mass spectral fragments (10). Rearrangement reactions are denoted by (r). Many of the m/e positions in the list are those that would be expected, especially the series 31, 45, 59, 73. Most of the m/e positions correspond to fragments containing oxygen, and are therefore not surprising. However, other m/e ratios such as 14, 27, 37, and 38 have no oxygen-containing fragments. These peaks appear to arise preferentially from the oxygen-containing compounds in the training set. They apparently are used by the learning machine to classify the compounds of the training set that cannot be classified on the basis of their oxygen-containing fragments alone.

Tables 4 and 5 show the results of applying the feature selection method to the problem of finding correlations between mass spectral features and

TABLE 3.   Probable Fragments of the 14 m/e Positions Correlating with Oxygen Presence

m/e	Fragments
14	CH_2
27	C_2H_3
31	CH_3O
37	C_3H
38	C_3H_2, C_2N
43	C_2H_3O, C_3H_7, CH_3N_2, C_2H_5N
45	C_2H_5O, $C_2H_7N(r)$
46	C_2H_6O
59	C_3H_7O, $C_2H_3O_2$, $C_2H_7N_2$, $C_2H_5NO(r)$
69	C_4H_5O, C_5H_9
73	C_4H_9O, $C_4H_{11}N(r)$
83	C_5H_7O, C_6H_{11}
100	$C_5H_{10}NO$, $C_6H_{14}N$, $C_6H_{12}O(r)$
135	$C_8H_7O_2$, $C_9H_{11}O$, C_8H_9NO

nitrogen presence. The two implementations were able to reduce the number of features to 43 and 42. The two lists of features had 37 m/e positions in common. These were supplied to the routines again, and each routine found one more ambiguous m/e position. Thus 36 features remained, and the learning machine correctly identified all the patterns of the training set and 93.3 and 94.3% of the prediction set using only these m/e positions.

Eighteen of the 36 selected features correlated with nitrogen presence, and they are listed in Table 6. The list contains more anomalies than for the oxygen case, but similar reasoning must be applicable. The m/e position of 14 is missing from the list, but 28 has been selected this time. Whether this indicates that the original samples had nitrogen impurities or that the learning machine used other information at m/e 28 is not clear, although the probability of nitrogen impurities for at least some samples is high.

Application of this feature selection routine based on computerized learning machines shows that the information relevant to the questions of oxygen presence or absence and nitrogen presence or absence is localized in relatively few m/e positions in low-resolution mass spectra. The routine employed is capable of learning to recognize all the members of its training set and a very high percentage of the prediction set while using only a fraction of the available features of the mass spectra. On closer examination the features selected by the routine depend on the fragmentation of these small organic molecules within the mass spectrometer, and show that some fragments not containing oxygen or nitrogen atoms are important nonetheless for determining their presence in the molecule.

TABLE 4. Training for Nitrogen Presence

m/e positions	+1 weight vector		−1 weight vector		Av % prediction
	Feedbacks	Predic-tion, %	Feedbacks	Predic-tion, %	
119	187	93.0	171	91.8	92.4
69	150	93.3	130	93.3	93.3
55	145	93.3	142	92.7	93.0
51	140	93.3	109	92.1	92.7
47	162	92.1	95	92.7	92.4
44	143	93.0	108	93.3	93.2
43	137	93.3	120	93.3	93.3
37	128	93.9	133	93.6	93.7
36	134	93.3	141	93.3	93.3

Av 93.0

	+	−	Total
Training set	38	262	300
Prediction set	43	287	330

TABLE 5. Training for Nitrogen Presence

m/e positions	+1 weight vector		−1 weight vector		Av % prediction
	Feedbacks	Predic-tion, %	Feedbacks	Predic-tion, %	
119	184	93.0	159	92.1	92.6
71	150	93.3	136	92.4	92.9
62	151	93.6	145	93.0	93.3
56	141	94.2	142	93.0	93.6
49	132	94.6	123	93.0	93.8
47	131	93.0	117	94.2	93.6
44	115	93.0	120	93.6	93.3
43	121	93.3	124	94.6	93.9
42	126	94.6	122	93.9	94.2
37	125	93.9	114	95.2	94.5
36	147	93.6	112	94.6	94.3

Av 93.6

The work just described was used as a basis on which to build in another investigation using the iterative least-squares training procedure. This training procedure was fully described in Chapter IV. It was used in conjunction with the weight-sign feature selection method and a larger set of low-resolution mass spectra as described in reference 11.

The data employed consisted of 450 low-resolution mass spectra from the API Research Project 44 tables as obtained from the United Kingdom Atomic Energy Authority on magnetic tape. Each spectrum consisted of numerous

TABLE 6. Probable Fragments of the 18 m/e Positions Correlating with Nitrogen Presence

m/e	Fragments	m/e	Fragments
15	CH_3	64	C_5H_4
27	C_2H_3	75	C_6H_3, $C_3H_7O_2$, $C_2H_7N_2O$
28	N_2, CH_2N, C_2H_4	76	CH_2NO_3, C_6H_4
30	CH_4N	93	C_6H_5O, C_7H_9, $C_6H_7N(r)$
41	C_2H_3N, C_3H_5	94	$C_6H_6O(r)$
44	CH_2NO, C_2H_6N, $C_2H_4O(r)$	96	C_7H_{12}
46	C_2H_6O, NO_2	106	C_7H_6O, C_7H_8N, $C_8H_{10}O(r)$
52	C_3H_2N, C_4H_4	116	C_8H_6N, C_9H_8, $C_6H_{14}NO$
60	C_2H_6NO, $C_2H_4O_2$	121	C_8H_9O, $C_7H_5O_2$

intensities listed in sequential order according to the m/e position. The lowest intensity reported was 0.01% of the largest peak in the spectrum. There were 132 m/e positions having at least 10 peaks in them out of the 450 spectra; therefore, 132 m/e positions was the upper limit on d, the dimensionality of the pattern vectors. The spectra were all of small organic molecules with molecular formulas $C_{1-10}H_{1-22}O_{0-4}N_{0-2}$. Each intensity was normalized according to the formula

$$I' = 10 \log_{10} (I1000)$$

in which I is the original intensity, and I' is the normalized intensity. (I' corresponds to x_{ij}, where i denotes the spectrum number, and j denotes the m/e index.) For storage purposes the values of I' were rounded off to the nearest integer. With the intensities in normalized form the new range of intensities become 10 to 50.

The set of normal equations to be used in the iterative least-squares method can be expressed compactly in matrix form. This system will have a unique nontrivial solution if the linear equations are independent. The rank of the coefficient matrix is the same as the dimension of the pattern space. Since the number of operations required to solve a system of linear equations for its solution increases as n^3, where n is the rank of the matrix, it is highly desirable to maintain the dimensionality of the matrix at a minimum.

The iterative least-squares method was applied to several test problems. The first problem was the determination of oxygen in small organic molecules, which had been studied previously. The problem was feature-selected by using the error correction feedback training method, and the number of features reduced from 132 to 31 with complete recognition and a prediction percentage of 93.9%.

For this problem Y_i was set equal to $+1$ if the ith spectrum contained oxygen, and to -1 if the ith spectrum did not contain oxygen. The weight vector components were initialized so that, if w_j had a value of p, w_{j+1} was assigned a value of $-p$. In this problem the value of w_1 was set equal to $+0.01$ or to -0.01. These initial values were found to be adequate in maintaining the values of the scalar products within the desired interval $(-2.5, +2.5)$. Minimization of the distance between the clusters has been found to facilitate convergence.

Table 7 lists the average values of several weight vectors after training under different conditions. If the average value is positive, that m/e position must correlate with oxygen presence. If, however, the average value is negative, then that m/e position must correlate with oxygen absence. The correlations obtained through the least-squares method were compared with the correlations obtained through the error correction feedback method. Twenty-two peaks agree in correlation, whereas nine disagree. The overall agreement of the correlations obtained through the use of two different methods supplies another clear indication as to which peaks correlate with oxygen presence and which peaks correlate with oxygen absence. The above agreement also suggests that there is some validity in the two methods used.

The rapid convergence of this previously feature-selected problem suggested that further feature selection could be performed. This was accom-

TABLE 7. Oxygen Presence Weight Vector Components

m/e	Average weight vector component	m/e	Average weight vector component
14	0.0208	46*	0.0081
15	−0.0250	52	−0.0178
17*	0.0275	59	0.0116
18*	−0.0003	63	−0.0215
24**	0.0007	67**	0.0102
25	−0.0003	69*	−0.0009
27	0.0049	70	−0.0026
30	−0.0183	73	0.0048
31	0.0211	83	0.0038
37	0.0042	84*	−0.0051
38**	−0.0035	86*	0.0002
39**	0.0195	91*	−0.0039
40	−0.0358	100	0.0084
43	0.0140	128	−0.0063
44	−0.0056	135*	0.0046
45	0.0080		

plished successfully, and the number of features was thereby reduced to 22. The mass positions discarded in going from 31 to 22 positions are marked with asterisks in Table 7.

The feature selection method consisted of using two different initial weight vectors. The training set was necessarily kept the same. The signs of the components of the weight vectors were compared after training. If they agreed, the mass position to which the components corresponded was retained but, if they disagreed, the mass position was eliminated. Feature selection resulted in no loss of recognition or prediction ability.

Table 8 presents the results obtained for the oxygen presence determination problem both with the original 31 peaks and after further feature selection. Recognition is nearly complete, and prediction percentage is of the order of 98%, regardless of the number of features employed.

This strengthens the argument that the features eliminated contributed little to the solution of the problem.

The lack of agreement of correlations for some peaks by using two different training methods suggested that the peaks whose correlations disagreed could actually be eliminated as in the feature selection routine. This was done, and the number of features was thereby reduced to 18. The features discarded in going from 22 to 18 positions are marked with two asterisks in Table 7. The results are given in Table 8. Recognition and prediction percentage do not suffer from such a removal. The correlations of the remaining 18 peaks are in perfect agreement with the corresponding correlations obtained from the error correction feedback method. This result strengthens the argument that some peaks can be used to identify lack of oxygen in molecules, whereas some other peaks identify the presence of oxygen.

The second test problem was determination of the presence or absence of nitrogen atoms in small organic molecules. Complete recognition was achieved, and therefore feature selection was possible. Thus the number of features was reduced from 132 to 43. Of these 43 features, those corresponding to the lowest 31 m/e positions were chosen.

TABLE 8 Oxygen Presence Problem

m/e Positions	Training set populations		Per cent recognition	Prediction set populations		Per cent prediction
	+	−		+	−	
31	42	108	99.3	26	74	98
	39	111	100.0	81	219	96
	43	107	98.7	77	223	98
22	43	107	99.3	77	223	98
18	43	107	99.3	77	223	98

As for oxygen, Y_i was set equal to $+1$ if the ith spectrum contained nitrogen, and to -1 if it did not contain nitrogen. The initial weight vector components were chosen as before.

Table 9 shows the results of the training procedure. The weight vectors were obtained under a variety of training conditions, including different training sets or different initial weight vectors. Averages of the weight vector components were found, and correlations with nitrogen presence or absence were determined. These were compared with the correlations obtained from the error correction feedback iterative method. Again, there is good agreement in correlations; 21 correlations agreed and only 10 disagreed.

Table 10 shows the results obtained for the nitrogen problem with the original 31 peaks and after further feature selection. The discarded features are marked with asterisks in Table 9. Again there is no loss of prediction or recognition ability. A prediction percentage of 96.7% was obtained for 31 features and of 98% for 21 features.

The correlations with nitrogen presence or absence resulting from different training methods were in good agreement. The peaks for which the correlations disagreed were removed, and training was continued with the remaining peaks. The discarded features are marked with two asterisks in Table 9. The results, as shown in Table 10, show that the peaks that were eliminated contributed little to the solution of the problem. A prediction of 98% was achieved by reducing the number of features to 14, a third of the number obtained by the error correction feedback method.

TABLE 9   Nitrogen Presence Weight Vector Components

m/e	Average weight vector component	m/e	Average weight vector component
12	−0.0220	53	−0.0228
13**	0.0340	55**	0.0254
15**	−0.0180	59*	0.0001
16**	0.0164	60*	0.0001
27	0.0002	64*	0.0005
28	0.0358	74	−0.0009
29	−0.0707	75*	−0.0001
30	0.0436	76	0.0137
31	−0.0056	85*	−0.0006
39*	−0.0009	86*	0.0002
41	−0.0101	87**	0.0001
43**	0.0001	91	−0.0004
44	0.0121	92*	−0.0003
45	0.0007	94*	0.0004
46**	−0.0001	96*	−0.0001
52	0.0216		

TABLE 10   Nitrogen Presence Problem

m/e Positions	Training set populations		Per cent recognition	Prediction set populations		Per cent prediction
	+	−		+	−	
31	12	138	99.3	20	280	95
	8	142	100	24	276	97
	9	141	99.7	23	277	94.7
	11	139	99.3	21	279	96.7
21	8	142	98.7	24	276	96
	17	133	98.7	15	285	97
	20	130	98.0	12	288	97
14	12	138	98.7	20	280	97

One of the problems encountered in the nitrogen problem was the population distribution. Over 90% of the spectra used were from compounds not containing nitrogen. Hence the decision hyperplane was shifted toward the cluster of nitrogen-containing compounds. As a result, recognition and prediction ability was much better for the nitrogen-lacking compounds than for the nitrogen-containing compounds. In order to alleviate this problem, the training sets were enriched with nitrogen-containing compounds by simply replacing some nitrogen-lacking compounds in the training set with nitrogen-containing compounds. The results show a much better placed decision hyperplane. The prediction for the nitrogen-containing cluster improved from 40 to 75%. Of course, the problem still remained, since the number of nitrogen-containing compounds was far too small. The best solution is to increase the data set by adding nitrogen-containing compounds, if possible.

The weight-sign feature selection method has also been used in the previously described investigation of SEP data (12). A total of 133 features was developed from the raw data, and weight-sign feature selection was used to reduce this number to 57 with essentially no loss in recognition. Ability to discriminate between one-component and two-component SEPs in the training set remained at approximately 96% with about 5 to 6% undecided. A further reduction in the number of features was accomplished by using the feature selection method described in the previous section of this chapter.

DISTANCE METRIC FEATURE SELECTION

A feature selection method has been developed (13) using a distance criterion to determine which features to drop.

Given a solution for a particular training set, each data point X_i in the set is characterized by a normal distance d_i from the decision surface. For a point on the proper side of the surface, this distance is given a positive sense and, for a point on the improper side of the surface, the distance is given a negative sense. The value of the surface half-thickness can now be defined as the minimum value of d_i occurring in the training set:

$$d_i = \pm \frac{|\mathbf{W} \cdot \mathbf{X}|}{|\mathbf{W}|}$$

$$t = \text{minimum } d_i$$

The value of t defined in this manner is the largest value that can be applied to the given decision surface and still represent a convergent solution, that is, all members of the training set are correctly classified.

In order to test the importance of a given descriptor, the feature selection routine temporarily omits it from the data set. Then values are obtained for all the d_i's, and t is determined. It is predicted here that the removal of an important descriptor would cause a considerable reduction in the value of t, whereas the removal of an extraneous descriptor would cause only a small variation in t. This procedure is repeated for each descriptor in the data set, and the resulting t's are compared. The algorithm assumes that the largest value of t corresponds to the descriptor that can most easily be eliminated, and it drops that descriptor from the data set. This procedure can then be repeated over and over, eliminating descriptors one at a time. However, it has been found that a superior performance is obtained if the weight vector is retrained periodically during the feature selection process.

A test of the algorithm was performed using a data set developed in an effort to simulate real infrared data. Each synthesized spectrum consisted of 128 descriptors. In preparing a single spectrum, all the descriptors were initially set equal to 100% transmittance. Next, a specified number of gaussian-shaped absorption peaks were coded onto the spectrum according to Beer's law. Each peak was centered about a randomly selected wavelength. The intensity of each peak and the full width at half-maximum were randomly selected within specified ranges. The data set consisted of 500 spectra each containing 20 randomly placed gaussian peaks with intensities between 40 and 80% and full widths at half-maximum randomly chosen between 0.1 and 0.4 μm. These parameters were selected in order to simulate real infrared spectra. The simulated infrared spectra appear remarkably similar to real spectra when plotted.

The spectra in the synthesized data set actually represent a collection of random backgrounds to which various features can be added in order to test the effectiveness of various training procedures. The advantage of such a data set, over a real data set, is that the user can create a carefully controlled training problem in which he knows in advance which descriptors are the important ones and what kind of information they contain.

A training set of 200 spectra was taken from the synthesized data set. To half of the spectra was added a peak of 30% intensity, 0.4 μm full width at half-maximum, at 4.4 μm. Of course, a fraction of the other synthesized spectra would also contain peaks in this region. In order to reduce the amount of calculation, only the first 50 descriptors were used (wavelengths between 2.0 and 6.9 μm). Weight vectors were developed that could discriminate between those spectra to which the 4.4-μm peak was added and those without such an addition.

The criteria used to decide when to call the training routine to retrain the weight vector were as follows: first, $t_{initial}$ was set equal to the value of t before the feature selection algorithm was called for the first time. On the first pass, the algorithm was used repeatedly until t became negative. Then the training routine was called with t set equal to $t_{initial}$. If convergence was obtained, the next pass was begun. If convergence was not reached, $t_{initial}$ was set to $\frac{1}{2} t_{initial}$, and the training routine was called again. Again, if convergence was obtained, the next pass was begun. If not, the training routine was called with t set equal to zero. If convergence was not reached here, calculations were terminated. Before each pass, the predictive ability of the decision surface was tested on a prediction set of 300 members (150 positive and 150 negative). The results are given in Table 11.

Passes 1 and 2 eliminated about two-thirds of the descriptors quite easily. Subsequent passes eliminated only one descriptor at a time. Initially, many

TABLE 11   Results of Feature Selection

Pass Number	$t_{initial}$	Descriptors Eliminated This Pass	Descriptors Remaining	Percent Prediction
0	2.63	—	50	98.51
1	2.63	23	27	97.55
2	1.31	11	16	97.23
3	1.31	1	15	97.24
4	1.31	1	14	97.90
5	1.31	1	13	97.89
6	1.31	1	12	97.51
7	1.31	1	11	98.59
8	0.66	1	10	97.24
9	0.66	1	9	97.25
10	0.66	1	8	96.25
11	0.66	1	7	96.92
12	0.66	0	7	—

components of the weight vector had values close to zero, but eliminating descriptors is equivalent to setting the corresponding component of the weight vector equal to zero, so dropping such a component would have little effect on the dot products and would not alter the decision surface significantly. After the first two passes eliminated 34 descriptors, the remaining descriptors, although not bearing any predictive significance, that is, they were not near the twenty-fifth descriptor, still had considerable magnitudes, and their removal would alter the surface enough to require the retraining of the weight vector.

At the end of the eighth pass, 10 descriptors remained, and the predictive ability was 97.2%. The remaining descriptors were: 2.0 (+), 2.2 (−), 3.0 (−), 4.0 (−), 4.3 (+), 4.4 (+), 4.5 (+), 5.4 (−), 5.6 (+), 6.2 (−), where the sign in parentheses shows whether the particular micrometer interval correlated with the presence of the added peak at 4.4 μm or not.

A second version of the feature selection routine was also employed in which two descriptors were eliminated at a time before checking whether $t < 0$ required retraining. The results were comparable to those of Table 11, however, a considerable savings in computer time was realized. The pairwise feature selection routine was able to reduce the number of descriptors down to eight with four passes. With only eight descriptors the predictive ability was 96.3%. The remaining descriptors were: 2.4 μm (−), 4.0 (−), 4.3 (+), 4.4 (+), 4.5 (+), 5.4 (−), 5.8 (+), and 6.2 (−). These correlations are quite similar to those mentioned above.

Finally, the weight-sign feature selection routine was applied to this data set for comparison. This routine involves training two weight vectors, one with all components initialized to +1 and the other with all components initialized to −1. After training, the signs of the individual components are compared, and those components for which the signs are different are dropped. The procedure is repeated until no further descriptors can be dropped. For the present problem the results obtained were that only 12 out of the 50 descriptors were eliminated, but predictive ability stayed high, with a value of 97.8% for the 38 descriptors remaining.

This feature selection routine utilizing t was next applied to a data set consisting of 500 infrared spectra of simple organic compounds of formulas $C_{3-10}H_{2-22}O_{0-3}N_{0-2}$. The first 500 solution infrared spectra listed in the Sadtler tables that satisfied this criterion were selected for the data set. Each spectrum was digitized at 0.1-μm intervals, from 2.0 to 14.7 μm, giving a total of 128 descriptors. The transmittances were read as accurately as possible to the nearest percent. If the strongest absorption in the spectrum was greater than 5% transmittance, that is, the absorption was weaker than one that would give 5% transmittance, the spectrum was normalized, assuming the validity of Beer's law, so that the strongest absorption was equal to 5% transmittance. Finally, all descriptors were scaled as integers ranging from 0 through 31, corresponding to 0 and 100% transmittance, respectively.

The data set of 500 real infrared spectra was used to develop weight vector maps for compounds of several chemical classes. In each case the following method was used to break down the data set. The number of positive members of the entire data set of 500 was determined. Two-thirds of the positive members were put in the training set along with twice as many randomly chosen negative members. The remaining spectra were all put into the prediction set.

For each chemical class the following procedure was followed. A training routine was used with a nonzero threshold. The feature selection routine described above was called repeatedly to eliminate pairs of descriptors. Whenever t became negative, $\mathbf{W}$ was retrained, and the predictive ability was determined. Three chemical classes were investigated: carboxylic acids, esters, and primary amines.

Carboxylic Acids

The training set consisted of 40 positive members and 80 negative members; the prediction set was split 21 ($+$) and 359 ($-$). The results of training and feature selection are summarized in Table 12. The predictive ability remained high at 94.5% when only 18 descriptors remained, but declined thereafter. With six descriptors, convergence was not obtained, and the routine was terminated.

Figure 2 shows a plot of the carboxylic acid weight vector of 34 components. As before, the weight vector components have been scaled arbi-

TABLE 12 Training and Feature Selection with Carboxylic Acids

Number of Descriptors	Number of Feedbacks	Percentage Prediction
128	753	95.9
34	272	95.6
22	487	93.5
18	409	94.5
16	1143	91.7
14	2049	90.7
12	1508	91.2
10	1842	92.6
8	2838	88.5
6	—	—

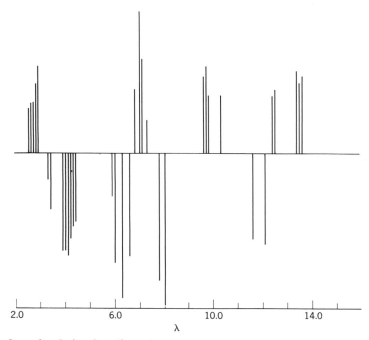

Figure 2.   Carboxylic acids weight vector map.

trarily and are plotted in the same orientation as for infrared absorption spectra. Several regions are of interest, working from left to right.

There is a definite negative correlation between carboxylic acids and the region between 2.4 and 2.8 μm. This is where free hydroxyl stretching is found. However, in undissociated acids the hydroxyls are usually hydrogen-bonded, which shifts the absorption to longer wavelengths. Thus this group of descriptors could be interpreted as ensuring that free hydroxyl groups are not classified as acids.

Eight descriptors, 3.2, 3.3, and 3.8 to 4.3, all correlate positively with carboxylic acids. It has been observed by spectroscopists that carboxylic acids generally exhibit a broad absorption peak in this region due to hydrogen-bonded hydroxyl stretching. Since such an absorption peak is relatively unique to carboxylic acids, this information was easily incorporated in the training of the binary pattern classifier.

Positive correlations are found for the 5.8- and 5.9-μm descriptors. This is the carbonyl stretching region.

The remainder of the descriptors have values that do not obviously correlate with some particular feature of carboxylic acids or are negatively correlated with carboxylic acids. These latter descriptor values are thought to be largely due to the particular negative members of the training set. Examination of the weight vector when it has only 18 components shows essentially the same correlations as those mentioned here.

Esters

The training set contained 40 (+) and 80 (−), and the prediction set contained 20 (+) and 360 (−). The results obtained are summarized in Table 13. The predictive ability fell slowly but steadily until no convergence could be obtained with eight descriptors remaining.

Figure 3 shows a plot of the ester weight vector with 36 descriptors. The positive correlations between esters and the descriptors for the 5.7-, 5.8-, 7.9-, 8.0-, 8.4-, 8.6-, and 8.9-µm regions are understandable, because of the standard ester stretch. The negative correlation for the descriptors 3.1, 3.2, and 3.5 to 3.8 may be due to the fact that these are especially useful in discriminating against C—H stretching vibrations of aldehydes and possibly the hydrogen-bonded hydroxyl stretching of carboxylic acids. The strong correlation between esters and the 2.0- to 2.7-µm descriptors is difficult to explain; it may be an instrumental artifact.

Primary Amines

The training set contained 38 (+) and 76 (−), and the prediction set contained 20 (+) and 366 (−). The results obtained are summarized in Table 14. The predictive ability remained steady with as few as 18 descriptors before falling.

Figure 4 shows a plot of the primary amine weight vector with 32 descriptors. A cluster of descriptors in the 3.6 to 4.0 µm region gives a positive

TABLE 13   Training and Feature Selection with Esters

Number of Descriptors	Number of Feedbacks	Percentage Prediction
128	720	96.6
36	458	95.5
24	604	94.4
14	541	93.5
12	319	93.2
10	516	91.5
8	—	—

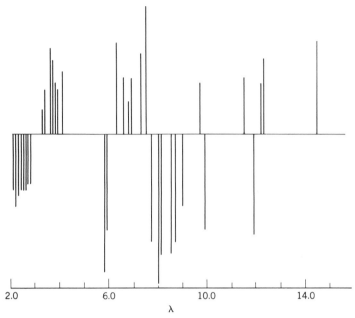

Figure 3.   Esters weight vector map.

correlation with primary amines, which is difficult to explain. This may be due to NH_4^+ modes which could be present if there were water in the samples. A group of descriptors in the 5.3 to 5.9 μm range were trained against characteristics of other classes of compounds. The 6.1- and 6.2-μm descriptors correlate with the N—H bending modes of primary amines. The positively correlated descriptors at longer wavelengths may correlate with known broad amine peaks in these ranges.

TABLE 14   Training and Feature Selection with Primary Amines

Number of Descriptors	Number of Feedbacks	Percentage Prediction
128	598	95.2
32	686	95.5
24	432	96.0
18	366	95.1
12	3560	92.3
10	3991	92.6
8	—	—

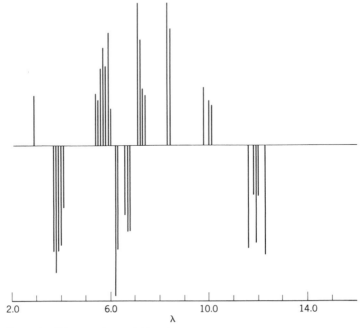

Figure 4. Primary amines weight vector map.

REFERENCES

1. J. T. Tou, *Pattern Recognition*, **1**, 3 (1968).
2. M. D. Levine, *Proc. IEEE*, **57**, 1391 (1969).
3. K. Fukunaga and W. L. G. Koontz, *IEEE Trans.* C-**19**, 311 (1970).
4. E. A. Patrick and F. P. Fischer, II, *IEEE Trans.*, IT-**15**, 577 (1969).
5. Many of the books listed in the Bibliography have chapters dealing with feature selection.
6. B. R. Kowalski, P. C. Jurs, T. L. Isenhour, and C. N. Reilly, *Anal. Chem.*, **41**, 695 (1969).
7. P. C. Jurs, B. R. Kowalski, T. L. Isenhour and C. N. Reilley, *Anal. Chem.*, **41**, 690 (1969).
8. P. C. Jurs, B. R. Kowalski, T. L. Isenhour, and C. N. Reilley, *Anal. Chem.*, **41**, 1949 (1969).
9. P. C. Jurs, *Anal. Chem.*, **42**, 1633 (1970).
10. F. W. McLafferty, *Advan. Chem. Ser.*, **40**, (1963).
11. L. Peitrantonio and P. C. Jurs, *Pattern Recognition*, **4**, 391 (1972).
12. L. B. Sybrandt and S. P. Perone, *Anal. Chem.*, **44**, 2331 (1972).
13. D. R. Preuss and P. C. Jurs, *Anal. Chem.*, **46**, 520 (1974).

CHAPTER VI

Further
Transformations

As discussed in Chapter I, the purpose of any pattern recognition procedure is to transform pattern space into classification space, in other words, to bring about transformations of the data that will place them into the desired categories. This can be viewed as a mapping from $(d + 1)$-dimensional space into a space of much lower dimensionality, often of only one dimension. Linear pattern classifiers dealing with each dimension independently were treated in considerable detail in earlier portions of this book. However, there remain many cases in which the pattern points are not linearly separable. If these cases are to be successfully treated, then it is necessary either to employ a higher-order decision surface or to use a transformation that will convert the data into a linearly separable set. (This statement is made on the assumption that the inseparability reflects the true nature of the data and is not due merely to inadequate descriptors, etc.)

The latter approach is very convenient in that it allows well-developed linear classification procedures to be applied. Therefore in this chapter we discuss some transformations that do not treat the dimensions independently. Several such transformations have been used to try to transform chemical data sets to make classification easier.

If the correct transformation were always known, of course, it would be a trivial matter to solve any problem by pattern recognition. The correct transformation is, in reality, rarely known, and the "game" becomes one of choosing a transformation that comes sufficiently close to allow the degree of accuracy required. A rigorous scheme for selecting the correct transformation is as yet unknown. Hence the intuition of the scientist must be applied. This approach finds analogies in many other areas, for example, the successful solution of differential equations often depends on the mathematician's ability to guess the correct transformation.

It should be realized that the linear learning machine is simply one method of finding the coefficients of a linear weighting function. If one assumes that an analytic function relates data to categories, a polynomial weighting function should perfectly classify all the data:

$$s = w_1x_1 + w_2x_2 + \cdots + w_{d+1}x_{d+1} + w_{11}x_1{}^2 + w_{22}x_2{}^2$$
$$+ \cdots + w_{12}x_1x_2 + w_{13}x_1x_3 + \cdots \quad (1)$$

The linear discriminant function can be considered just the first-order approximation of the polynomial function. It seems that the next logical step would be a second-order or quadric discriminant function. This function can be written in a form such that the individual terms are separated and there are seen to be d weights that are coefficients of the $x_j{}^2$ terms, d more weights that are coefficients of the x_j terms, $d(d-1)/2$ weights that are coefficients of the x_jx_k terms (cross terms, where $j \neq k$), and one weight that is not a coefficient but relates to the extra $(d+1)$st term. This can be written equivalently in the form

$$s = w_1f_1(X) + w_2f_2(X) + \cdots + w_mf_m(X) + w_{m+1} \quad (2)$$

in which $m = d + d + (d)(d-1)/2$. The important point to note is that the weights appear linearly in this expression. Thus a quadric discriminant function can be implemented by one of the following two methods:

$$X \rightarrow \begin{array}{c} \text{Quadric} \\ \text{discriminant} \\ \text{function} \end{array} \rightarrow \text{Classification}$$

$$X \rightarrow \begin{array}{c} \text{Quadric} \\ \text{preprocessor} \end{array} \rightarrow \begin{array}{c} \text{Linear} \\ \text{discriminant} \\ \text{function} \end{array} \rightarrow \text{Classification}$$

This result is quite general for a class of functions that have been termed Φ functions (1). The quadric function is a simple case of this general result which also holds for all polynomials, for example.

The observation that many complex discriminant functions can be implemented with a suitable preprocessor or transformer followed by a linear discriminant function is an important one. This had led to several studies of preprocessors, as described in the following discussion.

CROSS-TERM GENERATION

In almost all the early applications of pattern recognition techniques to mass spectrometry, linear TLUs were employed. These systems were linear in that they used mass spectral peaks independently of one another. However, the theory of mass spectrometry, as well as pattern classification considerations, suggests that second-order interactions (cross terms that consider relationships between peaks) could be used to advantage in making such classifications. Reference 2 reported the use of a similarity measure to develop two types of cross terms (intraset and interset cross terms) from low-resolution mass spectra. It was shown that the interset cross terms thereby derived had a high probability of being correlated with the molecular features that defined the categories of classification. The method was implemented with TLU pattern classifiers working with several sets of mass spectrometry data. The cross terms were shown to increase the power of the pattern classification systems by speeding up convergence and/or raising their predictive ability.

A mass spectrum in the form of a general, labeled graph is shown in Figure 1. In this symbolism the nodes $(v_1, v_2, \ldots, v_i, v_j, \ldots, v_p)$ represent the peaks that appear in the mass spectrum, and the arrows connecting the nodes (also called edges) represent the reaction pathways giving rise to the fragment ions. Thus v_1 in Figure 1 represents the parent, or molecular, ion from which all other ions are derived. For a graph with labeled nodes an adjacency matrix $A = [a_{ij}]$ can be formed. It is a square, symmetric, p by p matrix for a graph with p nodes. Its components are derived as follows: $a_{ij} = 1$ if node v_i is adjacent to node v_j in the graph, and $a_{ij} = 0$ otherwise. The p by p adjacency matrix is a complete, unambiguous representation of a graph with p nodes.

Many types of data (such as the mass spectral data of interest) can be expressed in vector form. That is, each mass spectrum in a data set can be expressed as $X = x_1, x_2, \ldots, x_p$, where each component of the vector corresponds to one peak in the spectrum so that x_{31} represents the intensity of the peak at $m/e = 31$. From a set of such vectors representing a data set, the following quantities can be calculated: b_{ii} is the number of vectors in the data set that contain a nonzero term x_i; b_{ij} denotes the number of vectors that have nonzero values for both x_i and x_j. For example, with a mass spectral data set containing 100 vectors, if x_{15} were nonzero for half of the vectors, $b_{15,15}$ would equal 50. A value of 40 for $b_{15,30}$ means that 40 vectors have peaks at both $m/e = 15$ and $m/e = 30$.

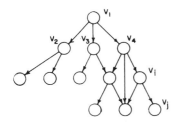

Figure 1   A generalized acyclic, directed graph.

These quantities are used to calculate the components c_{ij} of a term-term similarity matrix. If the vectors in the data set are all of length p, then the term-term similarity matrix is a square p by p matrix. Each element is computed as follows:

$$c_{ij} = \frac{b_{ij}}{b_{ii} + b_{jj} - b_{ij}} \tag{3}$$

Each element c_{ij} denotes the degree to which the ith and jth components of the collection of vectors are related. This measure of similarity is especially appealing for use with low-resolution mass spectra, because it deals explicitly with the locations of components in the vectors rather than their amplitudes.

The term-term similarity matrix constructed with equation 3 consists of elements c_{ij} between 0 and 1, where larger numbers indicate increased relatedness. This matrix can be converted to an adjacency matrix by applying a threshold T to each c_{ij} and setting $c_{ij} = 1$ if $c_{ij} > T$ and $c_{ij} = 0$ otherwise. The number of nonzero elements of the resulting adjacency matrix can be investigated as a function of threshold. Each 1 appearing in the adjacency matrix developed by the thresholding process corresponds to a single cross term which appears in the data set often enough to exceed the threshold value. Such cross terms might be useful features of the data for TLUs to use in classifying the data. Features developed in this manner clearly are intraset features, because the members of the entire set of vectors have been used together.

An approach for developing interset features is also based on the term-term similarity matrix. In this method the set of available data is split into the two subsets the pattern classifier will be trained to detect. Then, similarity matrices are formed using equation 3 for each of the two subsets, yielding matrices with terms c_{ij}^1 and c_{ij}^2. The absolute value of the difference is taken between these two matrices:

$$\Delta c_{ij} = |c_{ij}^1 - c_{ij}^2| \tag{4}$$

The resulting matrix has terms Δc_{ij}, where the magnitude of each element expresses the difference in the similarity measure of the cross terms formed from the ith and jth components in the two subsets of the data. This method of choosing interset features has been used with low-resolution mass spectra to choose cross terms to be incorporated into the set of features given to the pattern classification system.

For a set of 630 low-resolution mass spectra a term-term similarity matrix was developed using equation 3. Varying threshold values T were then applied to the similarity matrix. Curves a and b in Figure 2 show the number of cross terms (or edges) present in the data set as a function of threshold T. The uppermost labeling of the x axis was used in plotting curves a and b. The top curve refers to cross terms using all combinations of m/e positions, and the bottom curve refers to cross terms formed from two m/e positions that differ by more than 10 units, which gives a curve with the same general shape. With 119 m/e positions there are $(119)(118)/2 = 7021$ possible second-order cross terms, so the 1246 cross terms with thresholds of $T > 0.2$ are only a fraction of all the possibilities.

The curves on the right side of Figure 2 show the number of nodes in the largest cluster as a function of threshold. A cluster is defined as those nodes that are connected by an edge; any node in this type of cluster can be reached from any other node by traversing a series of edges. It is seen that the number of nodes in the largest cluster falls as a function of threshold T.

In order to find interest cross terms that are correlated with two chemical classes, the method described above for finding interset features was employed with mass spectra belonging to several specific chemical categories. A set of 450 mass spectra was divided into two subsets. Then the term-term similarity matrices for each of the two subsets of data were computed using equation 3, and finally the difference between them was taken using equation 4. The results of three implementations of this procedure are given in Table 1. They are for the following binary questions: (1) Is oxygen present

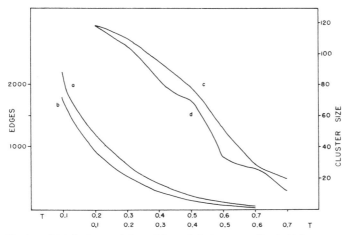

Figure 2. Number of total edges and sizes of largest clusters versus threshold for mass spectra with 119 m/e positions. (a) all m/e positions; (b) m/e positions differ by at least 10 units; (c) all m/e positions; (d) m/e positions differ by at least 10 units.

or absent? (2) Is the hydrogen number greater than 14 or not? (3) Is the hydrogen number more than twice the carbon number or not? The categories studied are typical of the types of information one would desire about a compound from its spectrum in a computerized learning machine.

The top portion of Table 1 shows the number of cross terms present for each subset over a range of thresholds. The different categories vary widely in the number of cross terms present for each threshold. For each category the number of cross terms given in row 6 of Table 1 was selected (these are the cross terms with the largest c_{ij} values, and the Δc_{ij} values were calculated. From the results of this procedure a small number of cross terms with the largest Δc_{ij} values were selected. The number selected for each category is given in row 7. The bottom portion of Table 1 gives profiles of the six sets of cross terms thus selected. The sizes and ranges of both c_{ij} values and Δc_{ij} values vary widely among the different classes. The largest Δc_{ij} values occur for the oxygen presence-absence test set, with average Δc_{ij} values of 0.74 and 0.66.

In order to see how well these cross terms correlated with the molecular features defining the subsets, the cross terms were injected into a pattern classifier using TLUs. All pattern classifier tests were made with a data set of 450 spectra each having 132 m/e positions. Three sets of tests were run with the three classifications used for Table 1. To start each test, a linear pattern classifier was trained using all 132 m/e positions. A randomly selected training set of 250 and a prediction set of 200 were used. The weight-sign feature selection routine was used to reduce the pattern space dimensionality by discarding m/e positions not helpful for the classification being performed. This was done with no degradation of convergence rate or predictive ability. Then the cross terms chosen as described above were inserted in the vacated positions, the classifier was trained again, and the predictive ability was checked. The results of this investigation are shown in Table 2. All tests were run in triplicate, with three different random training sets which are labeled A, B, and C. A dead zone of 50 was used during all training and for all prediction. The intensities of all peaks underwent a logarithmic transformation. Cross-term intensities were computed by multiplying the two m/e positions involved and taking the square root of the result to maintain normalization.

Table 2 presents the results of this procedure for oxygen presence determination. The pattern classifier is being trained to detect whether the compound whose mass spectrum is being classified contains oxygen in any functional form. The first column labels the triplicate runs by training set. The second and third columns give the number of linear m/e positions and the number of cross terms used by the pattern classifier. The fourth column

TABLE 1 *Interset Cross Terms*

Threshold	Oxygen Presence (120)	Oxygen Absence (165)	H > 14 (164)	H ≤ 14 (286)	2C/H < 1 (141)	2C/H ≥ 1 (309)
1.0	2	0	10	0	21	0
0.95	28	65	283	15	81	21
0.90	69	137	559	38	131	68
0.85	136	273		79	194	124
0.80	220	432		125	292	242
No. of cross terms investigated	220	220	200	200	292	242
No. of cross terms selected	13	14	11	10	10	11
Largest Δa_{ij}	0.769	0.688	0.498	0.652	0.551	0.367
Average Δa_{ij}	0.738	0.658	0.456	0.560	0.514	0.313
Smallest Δa_{ij}	0.701	0.613	0.413	0.393	0.501	0.301
Largest a_{ij}	0.886	0.912	0.987	0.903	0.932	0.852
Average a_{ij}	0.818	0.898	0.972	0.791	0.914	0.826
Smallest a_{ij}	0.805	0.881	0.968	0.753	0.901	0.814

TABLE 2 Oxygen Presence Training

Training	No. of m/e Positions	No. of Cross Terms	No. of Feedbacks	Convergence Improvement	Predictive Ability	Average Predictive Ability	No. Not Predicted	No. of Discarded Features
A	95	—	157/151		96.9/98.5	97.7	6/8	11
	95	27	79/89	0.54	95.4/95.4	95.4	3/3	19/1
B	95	—	185/164		98.4/97.9	98.1	10/9	12
	95	27	147/122	0.77	98.0/96.9	97.5	5/4	13/1
C	95	—	284/320		98.9/98.4	98.7	11/8	13
	95	27	173/167	0.58	99.5/99.0	99.3	5/10	49.0

Training Set		Training Set		Prediction Set	
		+	−	+	−
A		74	176	46	154
B		68	182	52	148
C		73	177	47	153

gives the number of feedbacks required to attain 100% recognition; the two figures each refer to one of the duplicate runs with different weight vector initialization (first all $+1$, and then all -1). The fifth column gives a measure of the improvement in convergence rate obtained when the cross terms are included. It equals the total number of feedbacks used during training with the cross terms, divided by the total number of feedbacks used during training with only linear terms. Thus a figure of 1.0 means the convergence rate was unchanged; a figure of 0.5 means that convergence was twice as fast with cross terms as without them. Column 6 gives the predictive abilities exhibited by the two pattern classifiers; column 7 gives their average. Column 8 gives the number of patterns out of 200 in the prediction set that were not classified because their dot products fell within the dead zone. Column 9 gives the number of features in the patterns that were discarded by the feature selection routine after training. (Features were discarded for which the corresponding weight vector components of the two weight vectors disagreed in sign.) For the training using cross terms, the two figures in column 9 refer to the number of linear terms and the number of cross terms discarded. The training and prediction set populations are given at the bottom of the table.

For oxygen presence training the linear feature extraction routine reduced the number of m/e positions from 132 to 95. With 95 m/e positions the pattern classifier quickly converged to perfect recognition and displayed predictive abilities of 97.7, 98.1, and 98.7%. With the addition of 27 cross terms (the same cross terms discussed in Table 1), the performance of the pattern classifiers changed. The convergence improvement figure ranges from 0.54 to 0.77 for the oxygen presence determination tests. This major increase in convergence rate demonstrates that the cross terms picked bear a strong correlation with the presence or absence of oxygen. The predictive ability was only slightly affected for training sets B and C, and it dropped slightly for A. Column 8 shows that in each case the pattern classifiers attempted more predictions when cross terms were used than when only linear terms were used. The ninth column shows that in each case more linear features of the patterns were discarded by the feature selection routine for training using cross terms than for linear training. Nearly all the cross terms were found to be useful by the feature selection routine.

Tests similar to the one shown in Table 2, but with cross terms selected by the experimenter or on the basis of c_{ij} values (intraset cross terms), showed that such cross terms were not helpful to the pattern classifier. Thus it has been shown that selection of interset cross terms on the basis of Δc_{ij} values yields second-order features that are useful to the pattern classifier.

FOURIER TRANSFORM (3)

The Fourier transform is most often thought of as relating the time domain to the frequency domain, for example, the positive part of the FT of a finite cosine wave is just a peak centered at the frequency of the wave. More generally the FT is just frequency per unit of X, where $f(X)$ is the function being transformed. [A discussion of the general theory and some applications of the Fourier transform can be found in the book by Bracewell (4).] In interferometry X represents distance, while in mass spectrometry X corresponds to mass/charge ratio. That is, the transform may be thought of as a frequency analysis of the original mass spectrum.

A second method of interpreting the transform is helpful in appreciating data transmission advantages in the Fourier domain. Each point in the Fourier domain is a weighted sum of all the points in the original domain, that is, the data at a given point (e.g., a mass position) in the original domain are spread out over the entire spectrum in the Fourier domain. This is referred to as the averaging property of the transform. Hence when errors occur or bits are lost during transmission of the data in the Fourier domain, there is a minimal effect on the original spectrum, whereas had the data been transmitted in the mass domain, loss of one bit could possibly result in a meaningless spectrum. It will be shown that this effect can be used to advantage in pattern classification as well.

It is convenient to introduce some notation at this point to facilitate discussion. The FT of real data is generally complex, having a real part originating from the cosine transform and an imaginary part originating from the sine transform. The FT of a function $f(X)$ is given by

$$G(v) = \int_{-\infty}^{\infty} f(X)e^{i2\pi vX} \, dX \tag{5}$$

or equivalently,

$$G(v) = \int_{-\infty}^{\infty} f(X) \cos 2\pi vX \, dX + i \int_{-\infty}^{\infty} f(X) \sin 2\pi vX \, dX$$

$$= G_c(v) + iG_s(v) \tag{6}$$

In practice $f(X)$ has finite limits and is set equal to zero beyond these limits so that the integration is over a finite interval. Two additional means of representing the data in the Fourier domain are easily calculated from equation 6. These are the phase spectrum defined by

$$\Phi(v) = \arctan \frac{G_s(v)}{G_c(v)} \tag{7}$$

and the intensity spectrum defined by

$$I(v) = [G_c(v)^2 + G_s(v)^2]^{1/2} \tag{8}$$

With respect to pattern classification, the transformed data are a different representation of the same information, which may make the implementation of a classifier simpler or more amenable to linear methods. The purpose here is to illustrate the use of the fast Fourier transform (FFT) as an aid to pattern classification of real data (low-resolution mass spectra) and to show how in practice the averaging property may be exploited for reduction of dimensionality.

The data used in this work consist of 630 low-resolution mass spectra taken from the API tables. This set contained 387 CH compounds and 243 CHO, CHN, or CHON compounds all of low molecular weight (less than 200 amu).

The Fourier transforms were calculated with a SHARE routine (SDA 3465) devised by Cooley. All 630 spectra were transformed and put on disk in the Triangle Universities Computation Center's IMB 360/75 system for fast retrieval. The spectra were handled identically by the training programs once they were transformed. That is, it made no difference in what domain the patterns were represented, since the training programs were general and required only that the data be presented as vectors.

Four different data sets were established, each consisting of 630 patterns. These were derived from (1) the cosine part of the transform (real part of G), (2) the sine part of the transform (imaginary part of G), (3) the phase spectra, and (4) the intensity spectra.

The FFT algorithm requires 2^N points, where N is a positive integer, for calculation of the transformation. Hence each mass spectrum of 200 mass positions was augmented to 256 mass positions by addending 56 zeros. Upon application of the FFT there resulted a set of 512 points, 256 each from the cosine and sine parts of the transform. The practical significance of this with real data is that the cosine part is even, whereas the sine part is odd; hence all information is contained in 128 points for either case. The other 128 points are easily generated by symmetry, hence are redundant.

The data consisted of the four sets of 630 patterns each, each pattern consisting of 128 components.

To determine whether or not the basic information was still present in the Fourier domain, pattern classifiers were developed on the transformed data and compared with those developed in the mass domain for the same questions. Some illustrative results are shown in Table 3. In each case the positive category consists of compounds with the indicated carbon/hydrogen ratio,

TABLE 3 Feasibility of Training and Prediction in the Fourier Domain Using C–H Questions

Positive Category	Feedbacks to Convergence					Prediction (% Correct)				
	Mass Spectra	Cosine Transform	Sine Transform	Intensity Spectra	Phase Spectra	Mass Spectra	Cosine Transform	Sine Transform	Intensity Spectra	Phase Spectra
C_nH_{2n+2}	30	40	145	286	161	97	93	96	96	79
C_nH_{2n}	107	358	364	2100	94	95	95	94	94	83
C_nH_{2n-2}	35	55	46	1380	115	96	95	96	88	89

and all other compounds make up the negative category. The data set consisted of the 387 CH compounds, 200 of these randomly chosen as a training set while the remaining 187 were used as a prediction set. The most important result is that the basic information is still easily obtained from the Fourier data. Convergence rate (as measured by the number of feedbacks necessary to attain complete convergence) and predictive ability (classification of unknown patterns) in the Fourier domain are comparable to that in the mass domain.

The usefulness of the trained weight vectors is measured in two ways in this work. (1) Attaining complete convergence, and thereby demonstrating linear separability, in a reasonable number of feedbacks is necessary if W is to be utilized in lieu of conventional data search and retrieval systems. (2) Accurate predictive performance of the trained weight vector is required if it is to be used as a pattern classifier for unknown patterns.

In previous work, 5 out of 43 categories of hydrocarbons, for which linear separability of low-resolution mass spectra was expected, were not found to be separable after allowing 2000 feedbacks for a training set of 200 compounds (5). The results of testing the Fourier domain for each of these categories are shown in Table 4 in comparison to the mass spectra treatment. The categories ethyl, n-propyl, and vinyl indicate compounds that have such groups, and a branch-point carbon is one with three or more carbon–carbon bonds. Note that by using phase spectra convergence was accomplished in every case, while no other form of the data was made to converge in any of the cases. Hence it is seen that the Fourier form of data may answer questions the original data could not answer—at least within the convergence limit applied. This satisfies the first criterion for use as a searching device. However, prediction is poor for all questions, hence the second criterion is not satisfied.

It is often desirable in problems of pattern classification and search and retrieval systems to reduce the dimensionality of the data. This greatly decreases both computation time and data storage requirements. The information in each mass position is spread over all dimensions in the Fourier domain by the averaging property mentioned above. Therefore some components can be arbitrarily set to zero or otherwise distorted in the Fourier domain, and the mass spectrum is still obtained upon inverse transformation, but with a higher noise level. (Therein lies one advantage of the transmission of images in the Fourier domain.) The problems of data collection and manipulation operations during which errors occur and the arbitrary omission of data are similar to the problems of noisy data transmission. Hence, owing to averaging properties, Fourier transformed patterns might be preferable to normal patterns such as mass spectra in some circumstances.

TABLE 4 Fourier Domain Results on Questions That Did Not Converge in Mass Domain

Positive Category	Feedbacks to Convergence					Prediction (% Correct)				
	Mass Spectra	Cosine Transform	Sine Transform	Cosine + Sine Transforms	Phase Spectra	Mass Spectra	Cosine Transform	Sine Transform	Cosine + Sine Transforms	Phase Spectra
$C = C$	>2000 [a]	>10000	>10000	>10000	160	78	73	75	72	75
Vinyl	>2000	>10000	>10000	>10000	199	80	79	76	65	74
Ethyl	>2000			>10000	645	73			62	67
n-Propyl	>2000			>10000	422	72			60	81
>2 Branch carbons	>2000			>10000	217	62			62	66

[a] Signifies that training was discontinued at the indicated number of feedbacks without having attained covergence.

Pattern classifiers were developed for mass spectra and the cosine (real) component of the Fourier transforms thereof, after reduction of dimensionality by several methods. The question was a carbon/hydrogen ratio of 1:2 versus all other CH compounds. The same training set of 200 compounds, consisting of 81 compounds with a carbon/hydrogen ratio of 1:2, was used in each case. The remaining 187 compounds, consisting of 74 with a carbon/hydrogen ratio of 1:2, were used in forming the prediction set. The results are summarized in Table 5. (Computation was terminated in each case when a set computation time was reached.) In every case the Fourier form of the data allows considerable reduction without serious increase in convergence time or decrease in predictive ability. The effect of data omission in the mass spectra varies, with the case in which dimensions are discarded on the basis of smallest average magnitude being the best method tried, as least as far as convergence rate to the final step of 16 dimensions is considered. Two of the three methods using Fourier data actually show slightly better predictive ability than the best mass spectra results. Hence it is seen that Fourier data can be arbitrarily reduced in dimensionality to a great extent before degradation becomes noticeable, whereas the original mass spectra may be so treated only when a logical criterion, such as the smallest average magnitude peak, is used.

A further study of the use of the Fourier transform for preprocessing of mass spectra has appeared (6). The data set employed consisted of 450 low-resolution mass spectra of 132 m/e positions each.

The sequence of computations applied to the original mass spectra is as follows. The original spectra have peaks up to m/e 200, so the pattern vectors to be transformed are 200-dimensional. The intensities of these peaks are logarithmically transformed. The Fourier transform of each spectrum is taken with the FFT algorithm (7), resulting in a 256-dimensional complex vector. Only the real part of the Fourier transform vector has been saved; because the 256 components of the real part of the Fourier transform vector exhibit symmetry about the midpoint of the vector, only the first half of the vector has been saved. Thus the Fourier transform pattern has 128 components, which is approximately the same as the original 132 m/e positions that were used. Of course, there are many more components with zero intensity in the mass spectral patterns than in the Fourier transform patterns.

The first study undertaken in this investigation was to perform feature selection on the Fourier transform spectra and to investigate the capabilities of the pattern classifiers as a function of the number of descriptors used. As has been pointed out, discarding some of the descriptors from the Fourier transform patterns is equivalent to losing some information about all the original mass spectra rather than all information about some of the patterns (8).

TABLE 5 Comparison of Dimension Reduction by various Methods in Fourier Domain with the Mass Domain

	Dimension	Highest Dimensions Omitted		Dimensions Omitted Randomly		Smallest Average Magnitude Dimensions Omitted	
		Feedbacks to Convergence	Prediction (% Correct)	Feedbacks to Convergence	Prediction (% Correct)	Feedbacks to Convergence	Prediction (% Correct)
Fourier Spectra, cosine part	128	75	95	75	95	75	95
	96	79	95	81	96	92	97
	64	162	98	103	95	147	98
	48	319	97	118	94	293	98
	32	2696	97	112	93	2088	97
	16	>8000	80	>8000	94	>8000	97
Mass Spectra	182	74	94	74	94	74	94
	96	116	96	142	98	74	94
	65	231	95	156	96	73	95
	48	>2666	90	781	94	82	95
	32			<4000	82	126	95
	16					575	94

Table 6 shows the results obtained when feature selection was performed on the Fourier transform spectra. The chemical question being trained for is whether the number of hydrogen atoms in the molecule is greater than twice the number of carbon atoms in the molecule or not. That is, compounds with more than twice as many hydrogens as carbons (e.g., alkanes, amines, etc.) comprise one category, and those for which $2(nC) \geq nH$ (e.g., alkenes, ketones, aromatics, etc.) comprise the other category. This is evidently a characteristic of a low-resolution mass spectrum that is not immediately apparent.

The first column of Table 6 labels the three parallel runs by the training set employed. Each training set consisted of 250 spectra randomly chosen from the overall data set of 450 spectra. The remaining 200 spectra in each case were used as the prediction set. Columns 2 to 5 give several measures of the performance of the pattern classifiers achieved before feature selection. Each Fourier transform pattern has 124 descriptors to start, because the first four descriptors, which have enormous values, do not contribute to the overall ease of solution of the problem. The training was performed with the threshold $Z = 10$; between 600 and 1000 feedbacks were required to obtain pattern classifiers that could correctly classify all the members of the training sets. Two binary pattern classifiers were trained for each training set; the exact number of feedbacks required for training in each case is given in the column 3 as two numbers separated by a slash, where the first number is the number of feedbacks for one weight vector and the second number is for the second weight vector. The fourth column gives the number of spectra in the prediction set that were not classified because the scalar fell in the dead zone. The fifth column gives the average percent prediction for the two binary pattern classifiers. It varies somewhat between training sets.

Columns 6 through 9 give the capabilities of pattern classifiers obtained after extensive feature selection. For the three training sets there remained only 76, 70, and 78 of the original 124 descriptors. The number of feedbacks needed to converge to perfect recognition of the training sets is approximately the same as in the previous case, or somewhat less. Predictive abilities for pattern classifiers working with smaller descriptor lists are substantially the same as for the complete Fourier transform spectra. Thus feature selection has not degraded the performance of the pattern classifiers, but has only speeded up the computations.

Table 7 presents a comparison of the capabilities of pattern classifiers that use mass spectra or Fourier transform spectra as their pattern vector inputs. The chemical classes being trained for are the same as for the previous table. The left half of the table shows that pattern classifiers trained with mass spectral patterns containing 111 feature-selected m/e positions converge to 100% recognition with 1200 to 2000 feedbacks and display predictive abilities

TABLE 6 *Feature Selection of Fourier Transform Patterns*

Training set	No. of descriptors	No. of feedbacks	Av No. not predicted	Av % prediction	No. of selected features	No. of feedbacks	Av No. not predicted	Av % prediction
A	124	575/752	21	95.3	76	627/629	23	95.5
B	124	727/670	25	93.4	70	678/817	21	93.6
C	124	952/1063	18	94.6	78	903/1003	17	94.5

TABLE 7 *Comparison of Properties of Pattern Classifiers Using Mass Spectra and Fourier Transform Spectra*

Training set	Mass spectra				Fourier transform spectra			
	m/e positions	No. of feedbacks	Av No. not predicted	Av % prediction	No. of descriptors	No. of feedbacks	Av No. not predicted	Av % prediction
A	111	1155/1666	17	95.1	76	627/629	23	95.5
B	111	1386/1451	19	93.2	70	678/817	21	93.6
C	111	1752/2020	19	93.5	78	903/1003	17	94.5

of 93 to 95% on complete unknowns. The right half of the table gives the performance of the pattern classifiers trained with Fourier transform spectra. The Fourier transform spectra consist of 76, 70, and 78 descriptors per pattern for the three training sets, Convergence occurs with 600 to 1000 feedbacks, approximately one-half as many as were necessary in the previous case. The predictive abilities achieved are better in each of the three cases; the increases are 0.4, 0.4, and 1.0%. Thus the capabilities of the classifiers dealing with the selected Fourier spectra compare favorably with the classifiers dealing with the original data.

The remainder of the investigation involved testing the reliability of TLUs trained with the original mass spectra and the Fourier transform spectra. It was shown that binary pattern classifiers trained with the Fourier transform spectra were more reliable.

The fast Fourier transform has been applied to the problem of NMR spectroscopy interpretation by pattern recognition methods (9). In this work the autocorrelation functions of simulated NMR spectra were employed as the data set. The autocorrelation removes the translational variance from the spectra and makes them more amenable to interpretation by pattern recognition methods. The autocorrelation function $A(x)$ of a function $F(t)$ is given by

$$A(x) = \int F(t)F(t + x) \, dt \tag{9}$$

The function $F(t)$ represents a continuous-wave NMR spectrum. The autocorrelation function can be approximated by a series and directly calculated from the defining equation. However, it is more computationally convenient to calculate the function by the following series of steps. (1) Take the Fourier transform of $F(t)$ to obtain the function $G(x)$. (2) Multiply $G(x)$ by its complex conjugate to form the power spectrum $|G(x)|^2$ of the original function. (3) Take the inverse Fourier transform of $|G(x)|^2$ to obtain the autocorrelation function. This three-step procedure can be conveniently carried out using the FFT algorithm. The possibility of using this approach for reduction of the size of the pattern vectors, that is, feature selection, was explored.

The study employed stick spectra calculated from input chemical shifts and coupling constants for which autocorrelation functions were calculated. The data set contained 634 patterns of 236 dimensions each. Several linear discriminant functions were trained on structural questions such as the presence or absence of the n-propyl group, and excellent results were reported.

The Hadamard transform has also been investigated for use as a preprocessor for mass spectral data (10). The Hadamard transform is analogous to the Fourier transform, although the Fourier transform decomposes functions into sinusoidal components whereas the Hadamard transform decomposes them into square-wave components. Results were presented for classification of hydrocarbons as C_6, C_7, or C_8 compounds from their low-resolution mass spectra after Hadamard transform preprocessing.

FACTOR ANALYSIS (11)

Another method of transformation is that known as factor analysis. The basis of this method is the diagonalization of a correlation matrix to find the eigenvalues of the matrix. The eigenvalues are used to rank the "importance" of the associated eigenvectors. The basic assumption is that the data and the categories into which they are placed are related through the variance.

Factor analysis has recently found interesting applications in the study of fundamental properties of solutes and stationary phases affecting gas chromatographic retention times (12, 13), in factors affecting solvent shifts in NMR (14–16), and in polarographic studies (17).

An objective of factor analysis is to find a set of variables fewer in number than the original set, which adequately describe or express the original set. What is "adequate" is determined by the experimenter. It may be sufficient to reproduce the data within experimental error, or less stringently, to reproduce major variations while ignoring smaller effects.

One begins with a matrix D of observations, such as a collection of mass spectra. The original experimental data are standardized by subtracting the mean intensity and dividing by the standard deviation for each mass position. From the standardized data, a correlation matrix C is calculated which relates the variation in each mass position to all the other mass positions.

$$C = D^T D \tag{10}$$

The m by m correlation matrix C can be transformed into a correlation matrix in an orthogonal coordinate system composed of linear combinations of the correlations in C, chosen so that the variance represented by C is distributed in as nonuniform a fashion as possible. In other words, the first coordinate axis, or linear combination, should contain the maximum variance possible on a single axis. The second linear combination should contain the second largest variance, while constrained to an axis orthogonal to the previous axis. Linear combinations expressing maximum variance are constructed until all the variance is accounted for. This orthogonal coordinate system is obtained by diagonalizing the correlation matrix to establish a set of eigenvalues E and a set of associated eigenvectors B:

$$B^{-1}CB = E \tag{11}$$

The square roots of the eigenvalues correspond to standard deviations, and the eigenvalues themselves to variances along the associated eigenvector axis. Thus the correlation matrix can be reproduced by rearranging equation 11 to yield

$$C = BEB^{-1} = BEB^T \tag{12}$$

Since B is an othogonal matrix, its inverse equals its transpose, and equation 12 is equivalent to

$$C = e_1 b_1 b_1{}^T + e_2 b_2 b_2{}^T + \cdots + e_m b_m b_m{}^T \tag{13}$$

However, as the eigenvalues e_i approach zero, little is gained by including the associated eigenvectors in the approximation of the correlation matrix. Needing fewer than m eigenvectors to reproduce the correlations in the original variables is the same as reducing the dimensionality of the space being studied, so that information compression results from reexpression of the variation by a more optimum coordinate system than the original variables.

Interpretation of the eigenvector solution of the correlation matrix in terms of importance to variation in intensity of mass spectral lines can be improved if the eigenvector axes are rotated in space to maximize the kurtosis (flatness) in the distribution of values within the eigenvectors. This is illustrated in Figure 3. Either coordinate system describes the arrangement of points in the figure, however, coordinate system B is more easily inter-

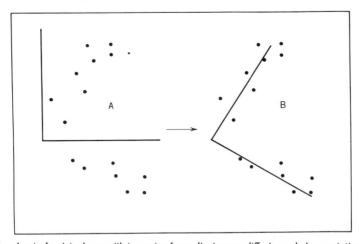

Figure 3.   A set of points shown with two sets of coordinate axes differing only by a rotation.

pretable in terms of the data, as vectors in the space will have large values for fewer data vectors.

In this study the coordinate system was rotated orthogonally to make the interpretation easier, and no attempt was made to rotate onto data vectors considered fundamentally important to mass spectral fragmentation patterns, such as ionization or appearance potentials. After rotation, the significant masses in each factor (a factor is an eigenvector scaled by the square root of its eigenvalue) were easily determined. For example, the largest coefficients in the twenty-sixth factor were for masses 45, 31, and 19. Because these masses are usually associated with oxygen, one would expect factor 26 to correlate with oxygen presence in the molecules studied. To test this and other factors for correlation with functional group presence in molecules, a method of ranking factors in relation to functional groups was used in which the dot product of a factor and each mass spectrum was calculated and the products P_{jk} summed over all mass spectra in a class. This was done for each factor using the equation

$$P_{jk} = \frac{1}{n} \sum_{i=1}^{n_k} F_j S_i \tag{14}$$

in which F_j is the jth factor, and S_i is in the ith standardized mass spectrum in class k. The standardized mass spectra were calculated from

$$S = \frac{I - \bar{I}}{\sigma} \tag{15}$$

in which I is the intensity of a peak, $\bar{I}$ is the mean intensity of the mass position, and σ is the standard deviation of the mass position. The results of equation 14 were ranked according to absolute magnitude, and the masses most significant in the highest ranking factors for each functional group were examined.

The results of applying equation 14 to seven functional groups are reported in Table 8. For carbonyls, two factors, 18 and 34, were of major and approximately equal importance, while three major factors, 26, 40, and 41, were found for hydroxyls. (Nonmajor are distinguished from major factors by a distinct decrease in the size of the dot product reported in column 3 of Table 8.) Factor 26 was also a significant factor in ethers, indicating that ether and hydroxyl fragmentation pathways are similar, while carbonyls follow a distinctly different route.

The variation in mass spectral intensity of 81 nitrogen compounds was related to two factors, 10 and 24, while four factors 10, 4, 19, and 37, were important to the 58 amines in the data set. Since factor 24 did not appear as a significant amine factor, an association with nitro compounds is indicated.

Saturated hydrocarbons are most significantly related to factor 16, while factors 23 and 25 are of less importance. Because the sums of products of factors and standardized spectra have been divided by the number of spectra in a class, the factors can be evaluated as to their relation to *any* functional group. Thus factor 26 is the most highly related of any factor to any functional group, and is twice as important to hydroxyls as to ethers. In fact, the data in column 3 indicate that all three hydroxyl factors are more related to hydroxyls than any of the ether factors are to ethers. An interpre-

TABLE 8 *Functional Group Relations for Factors for Seven Chemical Classes*

Functional Group	Related Factors	Relative Importance (Equation 14)[a]	Significant Mass Positions in Factor
Phenyl	15	−1.24	120, 105
	8	1.19	134, 133, 119
	21	−0.86	92, 91
	1	−0.65	117, 116, 90, 89
	31	0.58	129, 102
	32	0.50	122, 77
C=O	18	0.65	88, 61, 60
	34	−0.63	74, 73, 28
C—O—C	26	−0.80	73, 45, 31, 19
	9	−0.64	87, 75, 59, 47
OH	26	−1.60	73, 45, 31, 19
	40	−1.00	59, 33, 31
	41	−0.94	62, 32, 31
N	10	−0.51	123, 108, 18, 17
	13	−0.47	58
	24	0.40	76, 61, 46, 30, 17
NH_2, NH	10	−0.72	123, 108, 18, 17
	4	−0.62	98, 83, 70, 69, 56, 55, 42, 41, 39, 27
	19	0.55	72, 44
	37	0.53	96, 95, 94, 93
C_nH_{2n+2}	16	1.31	113, 99, 71, 57, 54
	23	−0.62	57
	25	−0.56	85, 84, 43

[a] Note that a negative value in this column does not necessarily imply that the masses associated with the factors are negatively correlated with the functional group. If the loadings themselves are negative, the double negative results in a positive correlation between the functional groups and the masses.

tation of this is that the fragmentation path of ethers is more disperse than that of hydroxyls, a not unusual conclusion given that the oxygen atom in ethers occurs in a more varied structural environment.

Factor 10 is more important to amines than to nitrogen compounds in general (0.72 versus 0.51), so that one associates the major mass positions of factor 10 with amines.

Once the important factors are known for a given functional group, the functional group mass positions are directly obtainable as those masses with large coefficients in the factor. These coefficients are the square roots of the variance in a mass position attributable to a given factor.

One can use the factor analysis approach to address the question: How much do functional groups affect mass spectra? To answer this question, the products of equation 14 for each of 35 functional group or structural properties were summed over the 42 factors. Because the mass spectra have been converted to standardized data, each spectra is a directed distance from the mean of all spectra. Factors are also directed distances from the mean of the data, obtained as orthogonal axes of maximum directed distance from the mean. Therefore the product of a spectrum and a factor is a measure of the similarity in direction and distance of the two vectors. By summing this measure over a class of spectra, one has an indication of the relation between the given class of compounds and an axis of variation in the data. Summation over all directions of variation is then an indication of a given functional group's ability to determine the fragmentation pattern of mass spectra. The results listed in Table 9 indicate which structural properties most influence the distribution of fragments in the 630 mass spectra studied.

The presence of the phenyl group influenced mass spectra most strongly. Nitrogen was the second most significant functional group, and the absence of all functional groups, that is, the saturated hydrocarbon structure was third. In other words, the presence of these groups shifted the spectra a significant distance from the mean spectrum of the data set, while the presence of a single methyl group did not cause the spectra to be significantly different from the mean spectrum of the data set.

The functional group with the greatest overall relation to the factor axes was the phenyl group, followed by three double bonds or less (which includes the phenyl group), amines, nitrogen, two double bonds or less, and saturated hydrocarbons.

A COMPLEX-VALUED NONLINEAR DISCRIMINANT FUNCTION (18)

The complex nonlinear discriminant function (CNDF) makes use of a generalized Walsh transform to construct the discriminant function (19). For a first-order generalized Walsh transform, in which the spectra are

TABLE 9 *Functional Group Factor Relations*

	Functional Group or Structural Parameter	Number of Spectra	Sum of Products of Equation 14
1	Phenyl	62	12.88
2	≥ 3 Double bonds	75	11.82
3	Amine	58	11.66
4	Nitrogen	81	9.60
5	≥ 2 Double bonds	108	9.48
6	C_nH_{2n+2}	89	9.83
7	Triple bond	42	8.65
8	≥ 2 Oxygen atoms	86	8.62
9	C—O—C	57	8.59
10	OH	33	8.25
11	Carbon atoms	86	7.71
12	C w/o H	103	7.35
13	>4 Methyl	106	7.15
14	≥ 9 Carbon atoms	184	6.48
15	Oxygen	174	6.29
16	C_nH_{2n}	154	6.24
17	Carbonyl	76	6.13
18	≥ 15 Hydrogen	205	5.69
19	≥ 2 Branch points	189	5.60
20	Vinyl	75	5.52
21	≥ 8 Carbon atoms	273	5.05
22	≥ 3 Methyl	220	4.86
23	≥ 1 Double bond	241	4.74
24	≥ 2 Ethyl	107	4.54
25	≥ 1 Branch point	309	4.30
26	*n*-Propyl	141	4.00
27	≥ 13 Hydrogen	313	3.97
28	≥ 7 Carbon	351	3.95
29	C w/o H	285	3.20
30	≥ 1 Ethyl	287	3.00
31	≥ 11 Hydrogens	405	2.84
32	≥ 6 Carbons	447	2.59
33	≥ 2 Methyls	420	2.41
34	≥ 9 Hydrogens	495	1.94
35	≥ 1 Methyl	543	1.12

allowed 50 integer intensities ranging from 0 to 49, each dimension of the
pattern or spectrum is transformed by the relation

$$T(I) = (e^{2\pi\sqrt{150}})^I \tag{16}$$

where I is the intensity at each mass position in the spectrum. Then $\Phi(x)$ is
the vector representing the transform of all intensities (dimensions) in the
mass spectrum:

$$\Phi(x) = T(I_1), T(I_2), \ldots, T(I_n) \tag{17}$$

A discriminant function using $\Phi(x)$ can be constructed having the form

$$F(x) = \theta + W^*\Phi(x) \tag{18}$$

W is simply the vector sum of all $\Phi(x)$ of the training spectra, given by

$$W = \frac{W_A}{a} - \frac{W_B}{b} \tag{19}$$

where a and b are the number of spectra in categories A and B, respectively,
and W_A and W_B are the weight vector components resulting from vector
summation of transformed spectra of compounds in categories A and B,
respectively. W^* is the conjugate transpose of W, obtained by changing the
sign of the imaginary part of the complex number. Since W is a vector,
transposing it involves no actual operation, but maintains mathematical
uniformity with matrix notation. θ is a constant related to the relative sizes
and variances of training set categories A and B. $F(x)$ is then a complex
number and is nonlinear with respect to the components of the mass spectrum. For prediction, compounds with positive real parts of $F(x)$ are classified
in one category, and those with negative $F(x)$ are put in the other.

From equation 18 it is seen that the discriminant function requires the
calculation of $\Phi(x)$ and W^*. Since the transformed spectral intensities can
assume only the 50 values given by equation 16, these intensities may be
calculated and stored for use in calculating the decision surface, rather than
recalculating them for each new spectrum, thereby greatly reducing computation time. W is calculated directly by equation 19.

The construction of the CNDF for mass spectra interpretation is implemented by taking two classes of compounds and transforming the mass
spectra of the compounds in each class according to equation 16. Considering one of the training set classes to be positive and the other negative, the
transformed spectra are summed algebraically to form a weight vector W,
which may then be used to predict the category of compounds not included
in the training set by using equation 18. For example, if the negative class
consists of compounds whose molecular formula is given by C_nH_{2n} and the
positive class consists of all other compounds, a compound whose calculated

$F(x)$ is less than zero is predicted to have the molecular formula C_nH_{2n}. In this manner predictions were made for 630 compounds using the categories listed in Table 10.

Overall prediction percentage was determined by training on all 630 compounds followed by subtracting the contribution to the weight vector of the compound to be predicted for and predicting the category of the compound. The compound's contribution to W was then added in again and the contribution of the next compound was subtracted. All 630 compounds were predicted for using the above method. The training set size was therefore 629 compounds. θ was determined by simultaneously predicting with a range of increments about zero added to the product of $W^*\Phi(x)$. These increments were plotted versus the percent predicted correctly, and the maximum of the curve was taken as the optimum θ. Table 10 lists the optimum θ for each question.

Table 10 shows the results of predictions based on the transformation of the original spectra and on sum-normalized transformed spectra. The sum normalization consisted of setting the sum of peak intensities for each compound equal to 100 and recalculating each peak accordingly. Any intensity greater than 49 was set equal to 49. The improvement in prediction resulting from setting the sum of intensities equal to 100 is interpreted as being due to the influence of individual compounds on the weight vector.

Table 10 includes the percent of total compounds in the larger category for each question. Since by always predicting that a compound belongs in the larger category one can predict at a level equal to the percent in the larger category, this value is a guide to whether the discriminant function has learned anything about the categories in question. The greater the difference in prediction percentage and percent in larger category, the better able the discriminant function is to differentiate two classes of compounds. Thus while the CNDF was not able to improve much on prediction of the carbonyl functional group, 87.9 versus 88.7%, it was significantly higher on questions such as detection and number of double bonds.

The percent predicted correctly for a given question can be used as an indication of the confidence to be placed in an individual prediction. For example, a prediction of the presence of a phenyl group (96.8%) is more likely to be right than a prediction of the presence of nitrogen (90.3%). This can be further refined. Figure 4 is a plot of the distributions of two classes, C_nH_{2n} compounds versus non-C_nH_{2n} compounds, as a function of $F(x)$. One would have the least confidence in a prediction based on a value falling very close to the decision surface. As the distance between $F(x)$ and the decision surface increases, the more confidence one has in the prediction. This is particularly useful when a variety of predictions is being made for

TABLE 10 *Predictive Ability of CNDF*

	Cutoff	Positive Category[a]	Negative Category	Percent in Larger Category	Theta	Percent Prediction, No Normalization	Percent Prediction, Sum Normalization
Oxygen	1	456	174	72.4	−0.622	87.6	88.3
	2	544	86	86.4	−0.326	90.0	90.0
Carbonyl	1	554	76	87.9	−1.614	87.9	88.7
Nitrogen	1	549	81	87.1	−1.111	88.3	90.3
Amine	1	572	58	90.8	−1.422	91.4	92.9
—C=C—	1	389	241	61.8	−0.429	80.0	82.5
	2	522	108	82.9	−1.466	94.9	95.2
	3	555	75	88.1	−2.014	98.3	89.3
C_nH_{2n}	—	476	154	75.6	−1.955	91.6	96.8
C_nH_{2n+2}	—	541	89	86.0	−1.777	95.9	95.6
Methyl	1	87	543	82.8	1.022	87.5	88.7

Ethyl	1	341	287	54.5	−0.148	71.4	77.1
	2	523	107	83.1	−0.340	86.5	86.0
Phenyl	1	568	62	90.2	−2.311	96.5	96.8
Carbon	5	105	525	83.4	0.458	91.6	92.1
	6	183	447	71.0	0.177	84.0	86.4
	7	279	351	55.7	0.088	77.3	85.2
	8	357	273	56.7	0.014	84.3	88.1
	9	446	184	70.8	0.192	86.7	85.4
	10	544	86	86.4	0.177	90.5	90.5
Hydrogen	9	135	495	78.6	0.311	85.6	87.8
	11	225	405	64.3	0.148	80.2	81.6
	13	317	313	50.3	−0.888	78.9	77.8
	15	425	203	67.5	−0.800	85.6	84.0
	17	501	129	79.5	−0.666	87.8	87.0
	19	554	76	88.0	−0.844	92.1	92.1

[a] Positive category contains compounds whose number of functional groups is less than the cutoff.

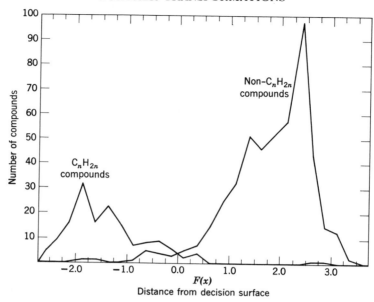

individual compounds. Conflicting predictions can be resolved by using the prediction that has the greatest probability of being correct, that is, the prediction having the greatest distance between $F(x)$ and the decision surface. This of course does not guarantee a correct decision, as the distribution shows.

The CNDF was also tested on the real, imaginary, and phase parts of Fourier transfom mass spectra. The results were comparable except for the phase part, which was lower (1 to 8%) on most questions. However, the phase prediction percentage was 92.2 for nitrogen and 95.1 for amines, a significant improvement. The improvement in nitrogen questions seems understandable because of the "odd mass" effect of nitrogen, causing peaks resulting from fragments containing nitrogen to be out of phase with non-nitrogen compound peaks.

REFERENCES

1. N. J. Nilsson, *Learning Machines. Foundations of Trainable Pattern-Classifying Systems,* McGraw-Hill, New York, 1965.
2. P. C. Jurs. *Appl. Spectrosc.*, **25,** 483 (1971).
3. L. E. Wangen, N. M. Frew, T. L. Lsenhour, and P. C. Jurs. *Appl. Spectrosc.*, **25,** 203 (1971).

4. R. Bracewell, *The Fourier Transform and Its Application*, McGraw-Hill, New York, 1965.

5. P. C. Jurs, B. R. Kowalski, T. L. Isenhour, and C. N. Reilley, *Anal. Chem.*, **42**, 1387 (1970).

6. P. C. Jurs, *Anal. Chem.*, **43**, 1812, (1971).

7. J. W. Cooley and J. W. Tukey, *Math. Comput.*, **19**, 297 (1965).

8. P. I. Boulton, E. J. Davison, and G. R. Lang, *Symposium on Feature Extraction and Selection in Pattern Recognition*, Argonne National Laboratory, Argonne, Ill., October 1970.

9. B. R. Kowalski and C. A. Reilley, *J. Phys. Chem.*, **75**, 1402 (1971).

10. B. R. Kowalski and C. F. Bender, *Anal. Chem.*, **45**, 2234 (1973).

11. J. B. Justice, Ph.D. Thesis, Department of Chemistry, University of North Carolina, Chapel Hill, N. C., 1974.

12. P. M. Weiner and D. G. Howery, *Anal. Chem.*, **44**, 1189 (1972).

13. P. M. Weiner and D. G. Howery, *Can. J. Chem.*, **50**, 448 (1972).

14. P. M. Weiner, E. R. Malinowski, and A. R. Levinstone, *J. Phys. Chem.*, **74**, 4537 (1970).

15. P. M. Weiner and E. R. Malinowski, *J. Phys. Chem.*, **75**, 1207 (1971).

16. P. M. Weiner and E. R. Malinowski, *J. Phys. Chem.*, **75**, 3160 (1971).

17. D. G. Howery, *Bull. Chem. Soc. Jap.*, **45**, 2643 (1972).

18. J. B. Justice, Jr., D. N. Anderson, T. L. Isenhour, and J. C. Marshall, *Anal. Chem.*, **44**, 2087 (1972).

19. Y. Uesaka, *IEEE Trans.*, **SMC-1**, 194 (1971).

CHAPTER VII

Molecular Structure to Properties

The chemical data interpretation tasks described to this point all involve the elucidation of chemical structure from observable data. The chemical data employed have been spectroscopic data with a few exceptions.

Now we turn to the reverse problem: the determination of physical properties of compounds directly from their molecular structure. This problem too can be approached using the same general pattern recognition techniques that have been discussed in detail. This approach depends on being able to codify the molecular structures of the compounds of interest in the pattern vector format necessary for compatability with TLU formalism. Thus this work leads directly into the field of chemical structure information handling.

Several types of physical properties of compounds have been studied in this manner. The most extensively studied is the generation of simulated mass spectra of small organic molecules directly from their suitable encoded molecular structures.

MASS SPECTRUM GENERATION (1–3)

Prediction of mass spectra by calculation has been approached in two ways: quasiequilibrium theory (QET), and heuristic DENDRAL, a computer program. Quasiequilibrium theory was developed from the observation that fragmentation processes leading to the formation of mass spectra could be viewed as rate processes similar to those occurring in ordinary chemical reactions. Models have been developed with varying degrees of sophistication, which are capable of calculating entire mass spectra for simple molecules (4–6). The first principal assumption of QET is that the molecular

136

processes leading to the formation of a mass spectrum consist of a series of competing, consecutive unimolecular decomposition reactions of excited parent molecular ions. The second assumption is that the rate constants for each of these reactions can be calculated using absolute reaction rate theory. Since the rates are functions of the internal energy of the reacting ion, energy-transfer functions are required to express the distribution of internal energy. Other necessary information includes reaction pathways, activation energies, parameters of the activated complexes, and information on electronic states. Curves are constructed giving the relative abundance of various ions as a function of energy; QET has been successfully applied to hydrocarbons and simple monofunctional compounds.

A portion of the computer program known as heuristic DENDRAL is capable of predicting the major features of the mass spectra of acyclic organic molecules (7). This subprogram, PREDICTOR, contains the (empirically obtained) basic theory of mass spectral fragmentation pathways. In its present form, PREDICTOR can handle aliphatic ketones, ethers, amines, and the general solution of aliphatic compounds of the form $C_nH_{2n+v}X$, where X represents oxygen, nitrogen, or sulfur, and v is the valence of X. The decisions made by the program are derived from experienced mass spectroscopists' knowledge.

The study was concerned with the generation of low-resolution mass spectra or organic molecules utilizing pattern recognition techniques and the molecular structure. Each binary pattern classifier was trained to predict the presence or absence of a mass spectral peak in each of 60 m/e positions. In addition, the system provided some intensity information for 11 of these m/e positions.

The data set was taken from a collection of mass spectra on magnetic tape available from the Mass Spectrometry Data Center, Atomic Weapons Research Establishment, United Kindgom Atomic Energy Authority. A section of the tape contains 2261 spectra from the API Research Project 44. A collection of 600 small organic molecules was chosen from this section along with their respective low-resolution mass spectra. The molecules were of the formulas $C_{3-10}H_{2-22}O_{0-4}N_{0-2}$. The digitized intensities ranged from 0.01 to 99.99% in each spectrum normalized to the most intense peak in the spectrum. The spectra were normalized so that the total ion current for each spectrum was equal. Then, an intensity of 9.5% for a particular peak means that the peak's intensity represents 9.5% of the total ion current in the spectrum.

Normally, the 600 compounds are input with a training set of 150 compounds chosen by a random number generation subroutine. The remainder of the 600 compounds are used as the prediction set.

Fragmentation Coding

The first requirement for implementing the binary pattern classifier is an adequate representation of the patterns (molecular structures) in computer-compatible form. Several of methods for representing chemical structures have been developed. Several reviews have described the various techniques and their advantages (e.g., 8). Obviously, the more unique the representation, the more likely is the possibility of success in applying binary pattern classifiers, and the better the simulation of mass spectra.

The technique of fragmentation coding was used. This method consists of representing a compound as a composite of its predominant structural fragments and their relationships. These features are then assigned numerical descriptors. Its advantages are that it is simple to learn, each to understand, yields a linear descriptor list which is immediately computer-compatible, and requires a moderate amount of computer storage. However, the simplicity of this method is to some degree offset by a loss in the complete description of a molecular structure. Fragmentation techniques do not normally indicate which fragments are bonded to one another, or what atom of a fragment is bonded to another. Information is lost in terms of the geometry and the stereochemistry of the molecule.

Several steps are followed in implementing the binary pattern classifier. Initially, a molecule is chosen from the data set. Second, from its three-dimensional structure, the chemist draws by hand a two-dimensional structural picture. As a third step, the two-dimensional diagram is checked against a previously chosen descriptor list. After all the compounds have been encoded as pattern vectors, and learning machine method is applied. If the patterns in the training set are linearly separable, a feature selection routine is used which reduces the number of descriptors necessary for linear separability.

Steps 2 and 3 can, in principle, be replaced. An alternative method of representing molecular structures such as linear notations (e.g., Wiswesser line notation) or graphical notations can be used. The descriptor list developed through fragmentation coding is not necessarily a unique representation of all molecules, and is biased toward the needs of this data set.

The 61 descriptors used in this study are listed in Table 1. The first column contains the names of the descriptors, most of which are self-explanatory. Several, however, need some explanation and are defined as follows. Largest clump refers to the largest number aggregate of carbon atoms bonded together. Largest cycle refers to the largest number of atoms (carbon, oxygen, or nitrogen), that can be traversed to complete a cycle (ring), but going through each atom only once, (e.g., naphthalene has a largest cycle of 10 atoms). Smallest cycle is the smallest number of atoms that can be traversed in a cycle, going through each atom only once, (e.g., naphthalene has a smallest cycle of six atoms). Number of rings plus double bonds is calculated from the following formula (9): the number of carbon atoms plus the number of hydrogen atoms divided by two minus the number of nitrogen atoms divided by two plus one, or $C + H/2 - N/2 + 1$. Ether, ketone, and alcohol are defined in the conventional manner, but combinations in one molecule are not allowed (e.g., to be considered an alcohol, a molecule can have only hydroxyl functional groups). Ester, carboxylic, and aldehyde refer to these groups being present in a compound, combinations being allowed (e.g., methylterephthalate contains both a carboxylic group and an ester group). Carbonyl presence means that there is a carbon–oxygen double bond in the compound. Oxygen linkage means that two carbon atoms are bonded together by an oxygen bridge. One benzene ring only refers to monocyclic molecules. For purposes of classification, a benzene ring is considered to have three double bonds. The branch-point carbon number is the number of carbon atoms in the compound that are bonded directly to at least three other carbon atoms. Methyl, ethyl, n-propyl, and n-butyl numbers are the numbers of each group that can be produced by a single bond rupture. The carbon w/o hydrogen category refers to quaternary carbon atoms that are not bonded to any hydrogen atoms. Two electron-donating groups ortho to each other refers to six-membered aromatic rings that contain two or more substituents of which two are in a position ortho to each other. The other substituent descriptors are described in a similar manner. The alpha-substitution category is defined as a methyl group ortho to a nitrogen atom in a ring. Gamma hydrogen refers to a hydrogen atom in the gamma position to a carbonyl group located in an acyclic compound. The other descriptors are as defined classically.

The descriptors are of two types, binary and numeric, as indicated in column 2 of Table 1. Binary descriptors can have only two values, corresponding to yes or no. Numeric descriptors can have values up to 202 (the molecular weight of $C_{10}H_{18}O_4$). Therefore it is necessary to normalize the values of the descriptors. The normalization constant by which each descriptor is multiplied is listed in column 3. After normalization the descriptors' values are in a more convenient range. All binary descriptors are normalized by a constant of value 5.

Column 4 lists any restrictions on the descriptors, or any special cases that may be encountered. For example, gamma hydrogen refers only to acyclic compounds.

Table 2 shows several examples of the compounds found in the data set. Under each compound is listed the numerical value assigned to each descriptor before normalization. Among other things it may be noted that

TABLE 1 *Molecular Structure Descriptors*

Descriptors	Type	Normalization constant	Restrictions
1. Molecular weight	N	0.05	...
2. Largest clump	N	1.00	...
3. Largest cycle	N	1.00	Not monocyclics
4. Carbon number	N	1.00	...
5. Hydrogen number	N	0.50	...
6. Oxygen number	N	3.00	...
7. Nitrogen number	N	5.00	...
8. No. of rings plus double bonds	N	1.00	...
9. Ether	B	5.00	...
10. Ester group	B	5.00	...
11. Ketone	B	5.00	...
12. Alcohol	B	5.00	...
13. Carbonyl group	B	5.00	...
14. Oxygen link	B	5.00	...
15. Hydroxyl group	B	5.00	...
16. Vinyl end group	B	5.00	...
17. Aromatic	B	5.00	...
18. Benzene ring presence	B	5.00	...
19. 1 benzene ring only	B	5.00	Monocyclic compounds only
20. Heteroatom in ring	B	5.00	...
21. Number of C=C	N	2.00	Benzene rings have three
22. Number of C≡C	N	5.00	...
23. Acyclic (no ring present)	B	5.00	...
24. Branch point carbon number	N	2.00	...
25. Number of clumps	N	3.00	...
26. Odd hydrogen number	B	5.00	...
27. No. of *n*-butyl groups	N	5.00	...
28. No. of methyl groups	N	2.00	...

#	Feature	Type	Value	Notes
29.	No. of ethyl groups	N	3.00	
30.	No. of n-propyl groups	N	5.00	
31.	No. of carbon w/o hydrogens	N	3.00	
32.	Carbon:hydrogen ratio 2n + 2	B	5.00	
33.	Carbon:hydrogen ratio 2n	B	5.00	
34.	Carbon:hydrogen ratio 2n − 2	B	5.00	
35.	Carbon:hydrogen ratio 2n − 6	B	5.00	
36.	Carbon:hydrogen ratio 2n − 4	B	5.00	
37.	No. of —CH₂— groups in a row	N	1.00	
38.	C=C—C—C—CH₃ ; methyl beta to a C=C	B	5.00	
39.	—C≡N group	B	5.00	
40.	—NO₂ group	B	5.00	
41.	—NH₂ group	B	5.00	
42.	N bonded to two or more carbons	B	5.00	
43.	Isopropyl presence	B	5.00	
44.	No. of rings	N	4.00	
45.	Size of monocyclic	N	1.00	
46.	Smallest cycle	N	1.00	Not monocyclics
47.	Fused rings	B	5.00	
48.	Alpha substitution	B	5.00	Nitrogen atom in a ring
49.	Gamma hydrogen	B	5.00	Acyclic compounds only
50.	Carboxylic acid group	B	5.00	
51.	Aldehyde group	B	5.00	
52.	2 electron donating groups, ortho	B	5.00	6-membered aromatic rings only
53.	2 electron donating groups, meta	B	5.00	6-membered aromatic rings only
54.	2 electron donating groups, para	B	5.00	6-membered aromatic rings only
55.	Nonfused rings	B	5.00	
56.	2 electron withdrawing groups, ortho	B	5.00	6-membered aromatic rings only
57.	2 electron withdrawing groups, meta	B	5.00	6-membered aromatic rings only
58.	2 electron withdrawing groups, para	B	5.00	6-membered aromatic rings only
59.	1 e⁻ donating − 1 e⁻ withdrawing, ortho	B	5.00	6-membered aromatic rings only
60.	1 e⁻ donating − 1 e⁻ withdrawing, meta	B	5.00	6-membered aromatic rings only
61.	1 e⁻ donating − 1 e⁻ withdrawing, para	B	5.00	6-membered aromatic rings only

141

TABLE 2 Descriptor Lists for Five Selected Compounds

Descriptor number	Compound Number				
	1	2	3	4	5
1	112	88	136	180	118
2	8	2	10	8	9
3	0	0	6	0	6
4	8	4	10	9	9
5	16	8	16	8	10
6	0	2	0	4	0
7	0	0	0	0	0
8	1	1	3	6	5
9	0	0	0	0	0
10	0	1	0	1	0
11	0	0	0	0	0
12	0	1	0	0	0
13	0	1	0	1	0
14	0	1	0	1	0
15	0	0	0	1	0
16	0	0	0	0	0
17	0	0	0	1	1
18	0	0	0	1	1
19	0	0	0	1	0
20	0	0	0	0	0
21	0	0	0	3	3
22	0	0	0	0	0
23	0	1	0	0	0
24	2	0	5	2	2
25	1	2	1	2	1
26	0	0	0	0	0
27	0	0	0	0	0

TABLE 2 continued

Compound number	API number	Structure	Chemical name
1	220		1, *cis*-2-Dimethylcyclohexane
2	326	$CH_3-C-O-CH_2-CH_3$ (with O double bond)	Ethyl acetate
3	466		1,7,7-Trimethyltricyclo-(2,2,1.0(2,6))-heptane (tricyclene)
4	1755		Methylterephthalate
5	1963		Cyclopropylb enzene

cis-trans isomerism is not to be found among the descriptors, nor are positions of functional groups of nonaromatic ring systems. The descriptors are chosen with the intent that the stuctures be divided into as many classes as possible. As will be shown, in many cases enough information is contained in the list as is necessary for linear separability.

Binary pattern classifiers were trained for the 60 m/e positions listed in column 1 of Table 3. Each of the binary pattern classifiers was trained to predict the presence or absence of a peak in its respective m/e position in the spectra of the compounds in its training set. To be considered present in a m/e position of a spectrum, a peak must have an intensity greater than a designated threshold value referred to as the intensity cutoff.

For all 60 m/e positions, weight vectors are developed for an intensity cutoff equal to 0.5% of the total ion current. A weight vector so trained can answer the following binary question: Does the compound under investigation contain a peak in this particular m/e position that has an intensity greater than 0.5%, or does it contain a peak with an intensity less than or equal to 0.5% of the total ion current?

In addition, for 11 m/e positions, two weight vectors each are developed for intensity cutoffs of 0.1 and 1.0% of the total ion current. The former predicts whether a peak has an intensity greater than 0.1% of the total ion current or less than or equal to 0.1% of the total ion current. The latter is able to answer a similar question for peaks with 1.0% of the total ion current. The choice of these 11 m/e positions was made to insure that there was an adequate number of compounds having a peak with an intensity greater than 1.0% of the total ion current in the training and prediction sets. A total of 82 binary pattern classifiers was trained.

Column 2 lists the intensity cutoffs used for each of the 82 TLUs. Impurities either in the samples or in the mass spectrometer itself, and the natural abundance of ^{13}C and ^{15}N, give rise to low-intensity peaks in the mass spectrum of a compound. The application of an intensity cutoff prevents the learning machine from utilizing this or other noise, to which it is quite sensitive, in its decision process. The lower limit of intensity that a peak can have is chosen as 0.1% of the total ion current of the spectrum in which it appears. All peaks with an intensity lower than this threshold are treated as not being present in the spectrum.

As mentioned, a training set of 150 compounds was chosen by a random number generation subroutine, and the remainder of the 600 was used as a prediction set of unknowns. Before training was initiated, all compounds in the data set of 600 whose molecular weight plus 1 was less than the m/e position were removed. It would be illogical to correlate the structure of a compound to an m/e position if the molecular weight of the compound were less than the m/e position being investigated. Since the lowest molecular weight

TABLE 3 *Feature Selection and Prediction*

(1) m/e	(2) Intensity cutoff (%)	(3) Descriptors remaining	(4) No. of feedbacks	(5) More populous category (%)	(6) Prediction (%)	(7) Columns (6)-(5)
29	0.1	15	270	89.7	91.3	1.6
	0.5	18	1246	80.7	92.0	11.3
	1.0	19	275	75.7	89.6	13.9
30	0.5	19	47	88.3	93.3	5.0
31	0.5	13	52	79.5	89.6	10.1
37	0.5	17	27	84.5	91.8	7.3
38	0.5	39	741	63.2	81.1	17.9
39	0.1	14	42	96.3	97.3	1.0
	0.5	14	20	93.7	96.2	2.5
	1.0	13	36	91.2	93.3	2.1
40	0.5	61	>2500	53.7	79.6	25.9
41	0.1	11	7	96.8	96.4	−0.4
	0.5	15	93	88.5	92.7	4.2
	1.0	25	83	82.7	93.8	11.1
42	0.1	18	>2500	87.5	94.0	6.5
	0.5	17	47	79.9	91.8	11.9
	1.0	61	>2500	56.2	72.1	15.9
43	0.1	19	294	85.1	89.8	4.7
	0.5	61	>2500	71.9	90.0	18.1
	1.0	61	>2500	64.7	84.0	19.3
44	0.5	61	>2500	70.5	76.5	6.0
45	0.5	15	42	81.5	90.8	9.3
46	0.5	13	33	96.1	95.3	−0.8
50	0.5	17	139	72.1	90.6	18.5
51	0.1	61	>2500	80.7	85.9	5.2
	0.5	61	>2500	56.1	93.3	37.2
	1.0	14	166	74.1	89.9	15.8
52	0.5	20	149	74.9	90.1	15.2
53	0.1	18	103	87.0	91.7	4.7
	0.5	26	173	56.4	86.3	29.9
	1.0	61	2349	63.6	86.7	23.1
54	0.5	61	>2500	70.8	82.2	11.4
55	0.1	35	574	79.2	81.7	2.5
	0.5	61	>2500	66.7	82.7	16.0
	1.0	30	1084	56.6	85.5	28.9
56	0.5	61	>2500	55.6	79.3	23.7
57	0.5	61	>2500	50.9	78.1	27.2
58	0.5	61	>2500	75.7	77.7	2.0
59	0.5	28	1058	87.2	83.3	−3.9
63	0.5	21	70	80.8	92.5	11.7
65	0.5	26	>2500	74.4	92.2	17.8

146

TABLE 3 continued

(1) m/e	(2) Intensity cutoff (%)	(3) Descriptors remaining	(4) No. of feedbacks	(5) More populous category (%)	(6) Prediction (%)	(7) Columns (6)-(5)
66	0.5	22	336	83.1	90.6	7.5
67	0.1	29	629	59.1	88.4	29.3
	0.5	29	1328	64.2	86.1	21.9
	1.0	34	>2500	74.8	85.4	10.6
68	0.5	61	>2500	76.1	86.3	10.2
69	0.5	61	>2500	56.1	87.0	30.9
70	0.5	61	>2500	61.2	81.4	20.2
71	0.5	61	>2500	76.0	81.3	5.3
73	0.5	18	251	90.1	89.0	−1.1
74	0.5	34	2265	90.1	86.6	−3.5
75	0.5	24	496	92.3	91.3	−1.0
76	0.5	12	12	94.1	96.3	2.2
77	0.1	15	56	52.0	89.4	37.4
	0.5	21	42	76.7	93.4	16.7
	1.0	18	31	82.5	90.1	7.6
78	0.5	10	161	84.2	94.0	9.8
79	0.1	28	873	54.1	86.7	32.6
	0.5	20	131	79.1	91.3	12.2
	1.0	16	104	87.6	89.9	2.3
80	0.5	18	250	92.9	93.9	1.0
81	0.5	61	>2500	80.6	91.3	10.7
82	0.5	25	>2500	84.1	87.8	3.7
83	0.5	61	>2500	74.8	88.8	14.0
84	0.5	61	>2500	72.8	75.8	3.0
85	0.5	61	>2500	86.4	88.5	2.1
91	0.5	11	491	83.2	92.0	8.8
95	0.5	13	141	88.2	93.2	5.0
97	0.5	61	>2500	84.2	87.3	3.1
98	0.5	61	>2500	78.0	76.2	−1.8
100	0.5	18	230	95.8	96.6	0.8
103	0.5	11	69	82.5	89.5	7.0
104	0.5	27	1365	85.1	90.1	5.0
105	0.5	8	58	85.0	92.1	7.1
106	0.5	15	39	91.1	95.5	4.4
115	0.5	18	68	81.8	93.5	11.7
119	0.5	14	26	83.5	95.4	11.9
120	0.5	18	31	83.9	94.7	10.8
121	0.5	13	27	89.6	92.1	2.5
127	0.5	6	5	93.5	93.2	−0.3
128	0.5	13	8	87.9	92.3	4.4
136	0.5	3	5	83.8	82.1	−1.7

in the data set is 40, all m/e positions below 41, inclusive, are unaffected by this procedure. Their training set size remains at 150, and their prediction set size at 450. At m/e position 83, the data set has been reduced from 600 to 500 compounds, to 400 at m/e position 98, to 302 at m/e position 106, and to 200 at m/e position 119. The data set consists of 149 compounds at m/e position 128, and only 68 compounds remain at m/e position 136. There is of course a corresponding proportional decrease in the training set and prediction set size from their original values as the data set size decreases. For example, the training set for m/e position 128 is reduced from 150 compounds to a total of 32, 4 with a peak whose intensity is greater than the cutoff of 0.5% of the total ion current, and 28 with a peak whose intensity is less than 0.5%. The prediction set is split into 14 compounds with a peak, and 103 without a peak, whose intensity is greater than 0.5% of the total ion current.

It is important to note that the intensity cutoff has an effect on the distribution of the training and prediction sets. The number of compounds that have a peak with an intensity greater than 0.1% of the total ion current for m/e position 29 is 137, and 13 have an intensity of less than 0.1%. With a cutoff of 0.5%, the number in the positive category is 121, and in the negative category 29. Similarly, for a cutoff of 1.0%, only 112 compounds are considered to have a peak in this m/e position, with 38 compounds considered to lack a peak in this m/e position. This trend was naturally observed for the training and prediction sets of the other m/e positions for which three intensity thresholds were applied.

Column 3 of Table 3 lists the number of descriptors that survived weight-sign feature selection by the learning machine. These descriptors were considered important to the presence of a peak in an m/e position, consistent with the intensity cutoff imposed on the data set. Despite the fact that in many cases the number of descriptors remaining is only a small fraction of the original 61, the weight vectors are able to retain essentially the same predictive ability as with all 61 descriptors.

The fourth column of Table 3 gives the number of feedbacks necessary for convergence; >2500 is indicated for those that have not been completely trained after 2500 iterations. Failure to train after a limited number of feedbacks does not prove linear inseparability. In some cases incompletely trained weight vectors display a high degree of predictive ability, as is observed for the TLU trained for m/e position 65. Linear inseparability does occur, however, in several cases tested; it is thought to be a consequence of incomplete description of the molecules. In response to this problem, additional descriptors were developed during the course of this investigation. Their appearance in Table 1 is due to obvious ambiguities in the description of the molecules that have been noted by the learning machine. Unfortunately, not all these ambiguities are readily apparent.

A simple test of the ability of the learning machine to classify unknown compounds is to compare its predictive ability with the percentage of compounds in the more populous category of an m/e position. If the predictive ability of the learning machine exceeds the success rate of always guessing the more populous category, the method is considered to have learned something about the relationship among the patterns and the category to which they are assigned. For example, the percentage of compounds having a peak with an intensity greater than a cutoff of 1.0% for m/e position 29 is 75.5. By always guessing that a compound contains a peak in this m/e position, one would make correct classifications 75.5% of the time. Similarly, for m/e position 128, the percentage of compounds that have a peak with an intensity greater than 0.5% of the total ion current is 12.1%. By guessing that all compounds in the data set do not contain a peak in this position, one would make correct classifications 87.9% of the time. Column 5 lists the percentage of compounds in the data set that occur in the more populous category, consistent with the intensity cutoff of the m/e position as listed in column 2.

Column 6 gives the predictive abilities of the weight vectors developed with the number of descriptors shown in column 3. For each of the m/e positions, three TLUs were trained with three different training sets. The resulting weight vector with the highest predictive ability has been chosen to be presented in column 6 for each m/e and each intensity cutoff. The average predictive ability for all 82 TLUs was 88.8%.

Column 7 contains the difference between the predictive ability of the learning machine as listed in column 6 and the percentage of compounds in the more populous category listed in column 5. In 73 cases the learning machine prediction exceeds the distribution of the peaks in that m/e position. The average amount by which the predictive ability exceeded the population of the more populous category was 10.4%.

Correlations of descriptors and m/e positions are of two types: those that are actual correlations of structural fragments and the m/e position, and those that are artifacts of the training sets. To minimize the latter, correlations were considered for all three training sets as a whole. After training and feature selection were finished for an m/e position, there remained three pairs of weight vectors, and thus three sets of descriptors for the m/e position. Since there were three training sets, not all the descriptors appeared on all three lists. A strong correlation would be implied if a descriptor were retained by all three lists. More importantly, if each of the pairs of weight vector components for these descriptors had the same sign, the correlation could be assumed to be real. (A pair of weight vectors for a particular training set has the same sign, although not necessarily the same value, for each descriptor, as a result of the feature selection method for reducing the number of descriptors.)

Table 4 shows the results of training and feature selection for m/e position 45 as an example. Although correlations are subject to influence by impurities and by isotope effects (i.e., fragments of m/e 43 and m/e 44 give rise to peaks of m/e 45 owing to the natural abundance of ^{13}C and ^{15}N), to some extent this is accounted for by the application of the intensity cutoff, in this case 0.5% of the total ion current. The training and feature selection using the three training sets yielded 15, 26, and 15 descriptors, corresponding to 35 different descriptors. Only five descriptors were retained by all three training sets, and these are listed in Table 4 along with their correlations. For the largest clump, ether, oxygen link, and hydroxyl presence descriptors, the correlations are negative, positive, positive, and positive, respectively. Such correlations refer to the fact that the sign of the weight vector components of these descriptors is the same for each of the three training sets. Two of the training sets gave positive correlations for the oxygen number descriptor, and one training set gave a negative correlation for the oxygen number descriptor; evidently this was due to an artifact of the training set employed.

The structural fragments that correspond to m/e 45 are CHO_2, C_2H_5O, and C_2H_7N. The first fragment is a carboxylic acid group. Since there are few carboxylic acids in the data set, there is little correlation with this fragment. The carboxylic acid descriptor did not survive the feature selection for all three training sets. The third fragment is likewise not associated with any of the descriptors in Table 4. It is caused by a rearrangement process and is not likely to be responsible for many of the peaks in this m/e position. Consequently, the learning machine has removed all nitrogen descriptors, considering them, like the carboxylic acid descriptor, unimportant to the solution of the problem. However, the ether, oxygen link, and hydroxyl presence descriptors relate well to the second fragment. According to McLafferty (10), this fragment originates from primary and secondary alcohol fragments, and an ether or oxygen link fragment as follows:

$HOCH_2CH_2$—Y   Y = —OR, —NR$_2$, —SR

$CH_3CH(OH)$—R   R = hydrocarbon moiety, generally aliphatic, but can be a hydrogen atom

CH_3OCH_2—R

The negative correlation of the largest clump descriptor is not as easily explained. It can be reasoned that, if there is a large number of carbon atoms linked together, then the probability of an oxygen atom or oxygen

TABLE 4   Results of Feature Selection for m/e Position 45ᵃ

Descriptors	Correlation
2. Largest clump	—
6. Oxygen number	+
9. Ether	+
14. Oxygen link	+
15. Hydroxyl presence (—OH)	+

ᵃ Descriptors retained by all three training sets, using an intensity cutoff of 0.5%.

fragment being present in a molecule is low. The fact that the oxygen number descriptor has both positive and negative correlations between training sets implies that the decision process employed by the learning machine is more complex than a simple acknowledgment of oxygen atoms in a compound. Rather, it is more important to note the arrangement of the oxygen atoms in the compound than their number.

As an overall test of the applicability of this method, 30 compounds were selected randomly out of the data set of 600. All 60 m/e positions were predicted for each compound using the 82 weight vectors developed. For 49 of the m/e positions, one can calculate a binary mass spectrum of the presence or absence of peaks in these m/e positions with an intensity greater than 0.5% of the total ion current. The use of three intensity cutoffs aids in determining a quantitative measure of the intensity of the peaks in the 11 other m/e positions. Each m/e position up to the m/e position one unit greater than the molecular weight of the compound is predicted for each of the 30 compounds. Table 5 indicates the applicability of the technique described in this study. The average predictive ability is seen to be 93%. This figure is slightly higher than the average predictive ability of all 82 TLUs, because some of the 30 randomly selected molecules were contained in the training sets of some of the TLUs. This predictive percentage of 93% demonstrates that the methods employed here are capable of a high predictive ability while using only a fraction of the descriptors available.

TABLE 5   Prediction of Entire Spectrum (30 Randomly
            Selected Compounds)

Total No. of errors	Errors on peak side	Errors on no peak side	Prediction (%)
154	85	69	92.9

Figures 1 to 6 show the predicted and real 11-peak mass spectra for six widely varying types of molecules out of the 30. The 11 m/e positions are labeled on the x axes, and the vertical axes for the real mass spectra are derived from the equation

$$I' = 10 \log (10000I)$$

in which I is the percentage of the total ion current for the spectrum due to the peak in question, and I' is the intensity plotted for the real spectra. A value of 30 on the y axis corresponds to an intensity of 0.1% of the total ion current, and values of 40 to 50 refer to intensities of 1.0 and 10.0% of the total ion current, respectively.

The values of the intensities for the m/e positions of the predicted mass spectra are assigned as follows. (Recall that the three TLUs were trained for cutoffs of 0.1, 0.5, and 1.0% of the total ion current, which correspond to

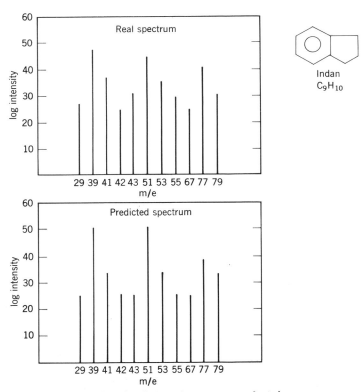

Figure 1.   Actual and predicted 11-peak mass spectrum for indan.

30, 37, and 40, respectively, on the logarithmic intensity scale being used in the figures.) The intensity given to a peak for which all three weight vectors yield a positive dot product is arbitrarily set at 50 (10% of the total ion current). If all three weight vectors yield negative dot products, the intensity is set to 25 (0.03%, which is probably in the noise). If the three weight vectors yield different scalars, the intensity is interpolated to 34 if the 30- and 37-cutoff weight vectors disagree, and 39 if the 37- and 40-cutoff weight vectors disagree.

The 33 TLUs that performed the classifications reported in the figures committed the following errors. The indan predictions contain two errors—m/e 77 and 79. The 2,4-dimethylpyrrole predictions are all correct. The isopropylacetate predictions contain two errors—m/e 42 and 55. (m/e 51, 53, 67, 77, and 79 are correctly predicted to be in the noise.) The ethylbenzene predictions contain one error at m/e 41. The 2,4-dimethylpentane and

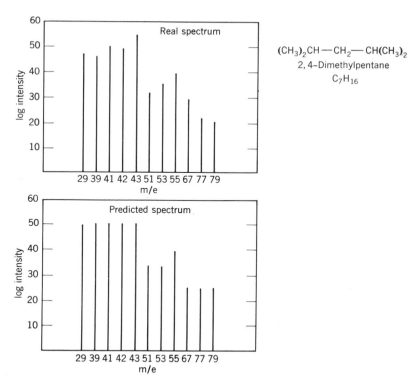

Figure 2.   *Actual and predicted 11-peak mass spectrum for 2,4-dimethylpentane.*

1,2,3- trimethylcyclohexane predictions are all correct. The six predicted mass spectra appear strikingly similar to the actual mass spectra. It should be noted that only the 11 peaks for which three TLUs were trained are shown for simplicity; 49 other m/e positions were also predicted in the overall test.

The study summarized in Tables 3 through 5 based on fragmentation coding of molecular structures demonstrated the possibility of predicting mass spectra directly from molecular structures. In order to improve the results, however, the deficiencies in that approach must be recognized and remedied. One problem was that the fragment code was an incomplete structural description; another problem was the one of balancing the populations of the yes and no categories for training. These problems and others have been addressed in studies to be described in the following discussion.

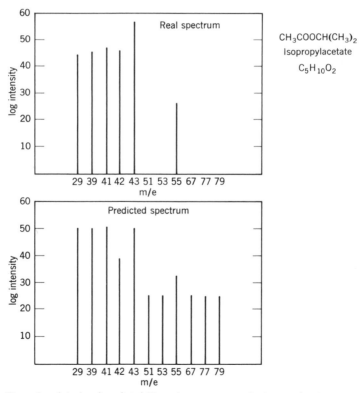

Figure 3. *Actual and predicted 11-peak mass spectrum for isopropylacetate.*

Multiple Features

The first problem attacked was that of developing descriptors that could take into account the simultaneous presence of several fragments. Such descriptors have been called multiple features.

Sections on feature extraction appeared earlier in this book, and the subject is not discussed again in general here.

The primary goal of this work was the pursuit of an improved description of molecular structures. The structural descriptors used previously primarily indicate the appearance of single structural fragments in a molecule. With the exception of a few descriptors, little attention has been paid to denoting combinations of structural fragments in a molecule. For example, descriptor 48, the alpha-substitution descriptor, indicates the appearance in a molecule of a methyl end group and a ring containing a nitrogen atom.

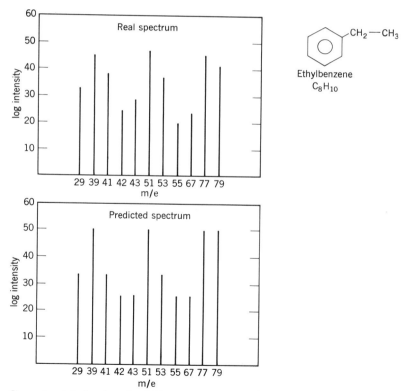

Figure 4.   Actual and predicted 11-peak mass spectrum for ethyl benzene.

It also describes the positional relationship of these two fragments. The methyl group is bonded to a carbon atom in the ring, which is directly bonded to the nitrogen atom. The automatic development of descriptors that incorporate several structural fragments into one is possible through the use of feature extraction techniques. This type of descriptor is referred to as a multiple feature. Multiple features, as developed from the structural fragments of the descriptor list, do not depict positional relationships. It may be possible to develop features that show positions of fragments relative to one another through the use of an alternative representation of the topology of a molecule, other than a descriptor list. The multiple features described in the following study indicate combinations of structural fragments in molecules, but not their positional relationships.

The feature extraction method used in this study is known as an attribute inclusion algorithm. Attribute inclusion describes the interrelationship of

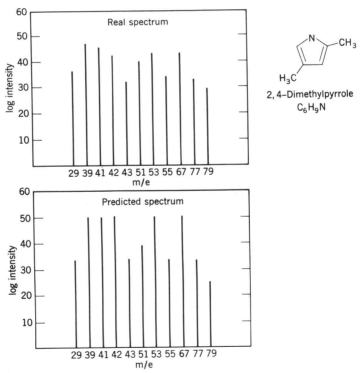

Figure 5. *Actual and predicted 11-peak mass spectrum for 2,4-dimethylpyrrole.*

attributes in a given set of patterns. In this case attribute is synonymous with descriptor. The molecular structures are represented by pattern vectors having attribute (descriptor) values as their components. The algorithm used is restricted to binary attributes (0 and 1). Therefore the pattern vectors presented to the algorithm are comprised of the 40 binary descriptors selected from the descriptor list of 61 structural fragments.

An attribute is included in another if whenever the first one is present in any pattern so is the second. Any two attributes satisfying the relation of inclusion belong together in a feature. Correspondingly, all attributes related by successive inclusions can be combined into a single feature. This feature constitutes a multiple feature or multiple descriptor. Therefore a set of features is desired that group together attributes correlated by mutual inclusion. Mathematically, attribute inclusion maps pattern vectors from attribute space into a feature space of lower dimensionality.

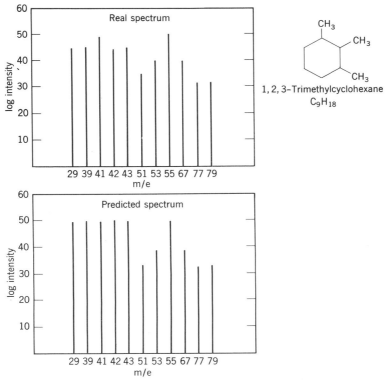

Figure 6.   Actual and predicted 11-peak mass spectrum for 1,2,3-trimethylcyclohexane.

The systematic way in which features are developed using the inclusion relation is described and applied to several examples of character reconstruction in an article by Abdali (11). The method is found to be sufficient for reconstructing patterns from the features it develops.

In Abdali's application, features are extracted from the entire data set of character patterns. Thus the features represent the common part of all the patterns containing it in the data set. Combinations of extracted features can then be used to reconstruct any of the patterns. A pattern is no longer described by its attributes, but by a lower-dimensional representation composed of features.

The attribute inclusion algorithm is applied to molecular structure data in a different manner than for the reconstruction of characters in the work of Abdali. Since this work involves pattern classification, rather than pattern reconstruction, it is more concerned with the development of multiple features that aid in the discrimination of the two pattern classes. Therefore each of the two classes is presented to the attribute inclusion algorithm separately, rather than together as one data set. In addition, the multiple features are not used as an alternative lower-dimensional representation of the original patterns. Rather, they are added to the original patterns, increasing the dimensionality of the pattern space. It is expected that the augmented patterns can be more easily classified into their respective categories than the original patterns.

The feature extraction algorithm is employed in the following way. After selection of a m/e position and an intensity cutoff, a training set of molecular structures is chosen and divided into two categories. Compounds that have a peak of sufficient intensity in this m/e position are stored as one class of patterns, referred to as category 1. Compounds that lack a peak of sufficient magnitude in this m/e position are stored as a second class of patterns in category 2. Each class of patterns is then presented to the feature extraction subroutine separately. In this manner a set of features is derived that is common to "peak" compounds, and another set is developed that is common to "no-peak" compounds.

The feature extraction subroutine is restricted to the 40 binary descriptors of the descriptor list. The patterns presented to the attribute inclusion algorithm, whether in category 1 or 2, are initially composed of 40 attributes or descriptors. The algorithm begins by removing any descriptors not present in any of the patterns of the category being investigated. During the course of development of the features, any descriptor that appears in exactly the same patterns as another descriptor is systematically removed. This condition is known as equality. The features constructed by the algorithm may contain one descriptor, or a number of descriptors. Features consisting of only one descriptor are normally the result of appearing in many of the patterns in the category. Multiple features, those that consist of more than one attribute (descriptor), are added to the end of the descriptor list. Those containing only one attribute are not added on to the end of the list because they already appear in the list. The algorithm is applied twice, once for each category. Any multiple features, from either of the two categories, become additions to the descriptor list beginning with descriptor 62.

An example of the feature extraction subroutine as applied to m/e position 45 is shown in Table 6. The training set consists of 25 compounds that have a peak with an intensity greater than a value of 0.5% of the total ion current, and 124 that have an intensity less than or equal to 0.5% of the total ion current. Both categories are represented by vectors having 40 components. The values of these components are either 1 or 0 (before normalization). After the 25 peak compounds are input to the feature extraction subroutine, 24 of the binary descriptors are removed from consideration of the inclusion relation because they do not appear in any of the 25 patterns. One attribute, descriptor 42, is removed because of the condition of equality. Ten multiple features are then calculated by the algorithm as shown. In this case, no features are developed that consist of only one descriptor. When the no-peak members of the training set are examined by the algorithm, seven attributes are removed because they do not appear in any of the 124 compounds of this set. From the remaining 33 attributes, attribute 14 is removed because of equality with another attribute. Thereafter 32 features are constructed, of which 16 are single features which are discarded because they already appear on the list of 61 descriptors. Seventeen multiple features are retained.

The 10 features from the peak compounds of the training set and the 16 from the no-peak compounds are added to the end of the descriptor list, augmenting the pattern vectors by 26 components, to make a total of 88 descriptors for each molecular structure.

The entire set of patterns, both training and prediction sets, is examined for the appearance of multiple features. A multiple feature is present in a compound if each and every one of the descriptors in the feature appears in the molecule. For example, every compound, whether it contains a peak or not, is examined for the presence of descriptor 79. If a compound contains an oxygen link, a carbonyl group, an ester group, is acyclic, and has a hydrogen atom gamma to a carbonyl oxygen, corresponding to descriptors 10, 14, 49, 13, and 23 of multiple feature 79, then a value of 1 is placed in the seventy-ninth component of its pattern vector. If it lacks any of these descriptors, then it is given a value of 0 for the seventy-ninth component. The normalization constant for the multiple features is set at 5.0, the same as for the binary descriptors.

Peak members of training set	25
Initial no. of attributes	40
No. with zero population	24
Final no. of attributes	16

Attributes remaining
  26 42 51 10 16 50 49 12 13 43 15 33 9 14 32 23
Attribute 42 removed because of equality

Descriptor	Multiple features
62	26 42 43 23
63	51 10 49 13 33 14 23
64	10 49 13 33 14 23
65	16 33 9 14 23
66	50 13 15 33 23
67	49 13 33 23
68	12 15 32 23
69	43 23
70	9 14 23
71	32 23

No. peak members of training set	124
Initial no. of attributes	40
No. with zero population	7
Final no. of attributes	33

Attributes remaining
  12 40 59 15 39 41 51 55 48 10 11 14 54 36 53 52 20 49
  35 42 19 13 26 47 38 17 18 34 16 32 43 33 23
Attribute 14 removed because of equality

Descriptor	Multiple features
72	12 15 55 34
73	40 26 23
74	59 39 19 26 17 18
. . .	15
75	39 26
76	41 26
77	51 13 23
. . .	55
78	48 20 42 26 17
79	10 14 49 13 23
80	11 13 23
81	54 35 19 17 18
. . .	36
82	53 52 35 19 17 18
83	52 35 19 17 18
84	20 42 26
85	49 13 23
86	35 19 17 18
. . .	42
87	19 17 18
. . .	26
. . .	47
. . .	38
. . .	17
. . .	18
. . .	34
. . .	16
88	32 23
. . .	43
. . .	33
. . .	23

A series of m/e positions is tested using the attribute inclusion algorithm as explained above. The binary pattern classifiers use simple error correction feedback training for the m/e positions investigated. It is expected that this type of feature extraction will increase the predictive ability of the binary pattern classifiers as a result of the better description of molecular structures. Unexpectedly, it is an aid in reducing the convergence time for training binary pattern classifiers.

Table 7 compares the speed of convergence for eight m/e positions using all 61 descriptors versus that for all 61 descriptors and any multiple descriptors developed by the attribute inclusion algorithm. As in all previous work, a training set of 150 compounds is randomly chosen, and the remainder of

TABLE 7   Effect of Multiple Features on Convergence Rate of Binary Pattern Classifiers[a]

(1)	(2)	(3)	(4)	(5)	(6)	(7)	(8)
	Without multiple features			With multiple features			
m/e	No. of descriptors	Prediction (%)	No. of feedbacks	No. of descriptors + multiple features	Prediction (%)	No. of feedbacks	Columns 4/7
29	61	88.7	63	87	88.0	57	1.1
	61	88.4	77	87	89.1	51	1.5
31	61	88.9	113	90	90.2	81	1.4
	61	87.1	91	90	88.7	81	1.1
45	61	89.9	130	88	91.7	34	3.8
	61	92.0	74	88	93.3	44	1.7
59	61	82.4	2056	90	83.1	391	5.3
	61	82.8	535	90	82.1	277	1.9
67	61	85.4	1244	88	88.0	844	1.5
	61	84.9	1442	88	81.4	735	2.0
73	61	90.0	664	88	90.5	492	1.4
	61	86.3	379	88	84.5	383	1.0
77	61	91.6	137	89	90.8	119	1.2
	61	92.4	43	89	90.5	33	1.3
104	61	89.2	915	81	88.4	655	1.4
	61	87.1	464	81	87.1	331	1.4

[a] Intensity cutoff for each of the m/e positions is 0.5% of the total ion current.

the data set of 600 is used as a prediction set. Before the feature extraction subroutine is employed, all compounds are removed from the training and prediction sets whose molecular weight is not consistent with the m/e position under investigation. This process has been explained previously. The only difference between this work and previous work is that the number of descriptors is increased for a binary pattern classifier utilizing the feature extraction subroutine. The training and prediction sets as well as the intensity cutoff for a particular m/e position are identical for both the case using multiple features and the one not using multiple features.

The first column of Table 7 is a list of the eight m/e positions studied. The intensity cutoff is 0.5% of the total ion current for each of the m/e positions. In column 3 is the predictive ability of the binary pattern classifier, using all 61 descriptors, as shown in column 2. The upper number of the pair in column 3 corresponds to a $+1$ starting weight vector. The bottom number corresponds to the -1 starting weight vector. Column 4 contains the number of feedbacks necessary for complete recognition of the training set. In column 5 is the total number of descriptors used by the binary pattern classifier, including multiple descriptors. This value is equal to 61 plus the number of multiple descriptors constructed by the feature extraction subroutine for the particular m/e position. Columns 6 and 7 show, respectively, the predictive ability using multiple descriptors, and the number of feedbacks for convergence. In column 8 is the ratio of columns 4 and 7. It represents the increase in the rate of training resulting from the addition of multiple descriptors to the pattern vectors.

For m/e position 67, binary pattern classifiers can predict with an accuracy of 85.4 and 84.9% the appearance of a peak with an intensity greater than 0.5% of the total ion current, using 61 descriptors. Training the $+1$ and -1 weight vectors requires 1244 and 1442 feedbacks, respectively. With the addition of 27 multiple features, nearly a 50% increase in the dimensionality of the patterns, the binary pattern classifier predicts the appearance of a peak with 88.0 and 81.4% accuracy using 844 and 735 feedbacks, respectively. This is an increase in the convergence rate of 1.5 and 2.0 times that required without the aid of multiple descriptors.

On the average the rate of convergence for the eight m/e positions is 1.8 times faster if multiple descriptors are used to augment the original pattern vectors. This increase in dimensionality is normally in the range of 50%. Since the feature extraction subroutine is quick, the time required for its execution is offset by the savings involved in the increase in the training rate, the slow step in the classification technique.

To test the increase in predictive ability using multiple descriptors, the binary pattern classifiers of Table 7 are allowed to undergo feature selection.

Table 8 shows the results of feature selection for the eight m/e positions previously examined. Again, the same training and prediction sets are used for a m/e position whether feature extraction is utilized or not. The intensity cutoff is 0.5% of the total ion current for each of the m/e positions listed in column 1. Columns 2 and 3 show the number of descriptors and the predictive ability without multiple features. Columns 4 and 5 correspond to the number of descriptors and predictive ability of the binary pattern classifiers that employed feature extraction. In each case the predictive ability shown is that occurring after feature selection has removed all unnecessary descriptors.

Without using multiple features, m/e position 59 is feature-selected to 28 descriptors with a corresponding predictive ability of 83.3%. With 30 descriptors the binary pattern classifier that uses multiple features has a predictive ability of 84.7%.

The average increase in predictive ability is 0.6% with the aid of feature extraction. Although not shown in Table 8, the average number of multiple descriptors retained by the eight binary pattern classifiers after selection is five. Also not noted is the fact that training after consecutive feature selections is faster for cases that use multiple features.

Substructure Descriptors

In a further attempt to improve the descriptor set, another type of descriptor was added to the pattern vectors. These were substructure descriptors which

TABLE 8 Effect of Multiple Features on Predictive Ability after Feature Selection

m/e	Without multiple features		With multiple features	
	Descriptors remaining	Prediction (%)	No. of descriptors + multiple features	Prediction (%)
29	18	92.0	24	92.9
31	13	89.6	19	90.4
45	15	90.8	17	93.3
59	28	83.3	30	84.7
67	29	86.1	42	85.2
73	18	89.0	21	89.3
77	21	93.4	23	92.6
104	27	90.1	31	91.0

[a] Intensity cutoff for each m/e position is 0.5% of the total ion current.

described positional relationships in addition to the presence of structural features.

The problem treated was the prediction of the mass spectra of hydrocarbon molecules directly from a description of their molecular structures. A set of 377 hydrocarbons with molecular formulas $C_{3-10}H_{2-22}$ were used. In order to proceed with the development of the programs, two sets of data must be generated: (1) molecular structure descriptions in the required vector format; and (2) a set of answers, that is, the mass spectra corresponding to the structures.

To be input to the learning machine programs, the molecular structure descriptions must be in the vector format previously discussed. Of the multitude of ways to do this, a combination of two popular approaches has been chosen: fragment codes and substructural codes. The technique of fragmentation coding consists of representing a compound as a composite of its predominant structural fragments and their relationships. These features are then assigned numerical descriptors. Table 9 lists the fragments used in the present molecular structure descriptions. They are mostly self-explanatory; the remainder are defined as follows. Largest cycle refers to the largest number of atoms that can be traversed to complete a cycle (ring), but going through each atom only once, for example, napthalene has a largest cycle of 10. Smallest cycle is the least number of atoms that can be traversed in a cycle, for example, napthalene has a smallest cycle of six. Only compounds with at least two rings have nonzero values for the largest cycle and smallest cycle descriptors. One benzene ring only refers to monocyclic molecules. For purposes of classification a benzene ring is defined to have three double bonds. The branch-point carbon number is the number of carbon atoms in the compound bonded directly to at least three other carbons. Methyl, ethyl, n-propyl and n-butyl numbers are the numbers of each group that can be produced by a single bond rupture. There are 29 fragment descriptors.

The descriptors are of two types—binary and numeric— and are labeled in Table 9. Binary descriptors can have only two values corresponding to yes and no. In a pattern vector a 1 corresponds to the presence of a fragment, and a 0 corresponds to its absence in the structure being coded. Numeric descriptors can have values of up to 142 (the molecular weight of $C_{10}H_{22}$). Because of the wide variation in the ranges of the descriptors' values, it is necessary to normalize the values of the descriptors. The normalization constants are chosen to decrease the spread in the range of the descriptors over the data set. For example, the normalization constant for hydrogen number is 0.5, making its range 0 to 11. All the binary descriptors are normalized by a constant of value 5.

The remainder of the descriptors are the 26 substructure descriptors listed in Table 10. They are all binary descriptors. The entire data set thus consists of 377 hydrocarbons with their structures coded into 55-dimensional pattern vectors.

The set of mass spectra used was taken from a collection of data on magnetic tape available from the Mass Spectrometry Data Centre, Atomic Weapons Research Establishment, United Kingdom Atomic Energy Authority. The 377 hydrocarbon spectra used were taken from part of a tape containing 2261 spectra from the API Research Project 44. The digitized

TABLE 9   *Fragment Descriptors[a]*

1. Molecular weight	N
2. Largest cycle	N (Not monocyclics)
3. Carbon number	N
4. Hydrogen number	N
5. Number of rings and double bonds	N
6. Vinyl end group	B
7. Aromatic	B
8. Benzene ring presence	B
9. One benzene ring only	B
10. Number of $C=C$	N
11. Number of $C\equiv C$	N
12. Acryclic	B
13. Branch-point carbon number	N
14. Number of *n*-butyl groups	N
15. Number of methyl groups	N
16. Number of ethyl groups	N
17. Number of *n*-propyl groups	N
18. $H = 2C + 2$	B
19. $H = 2C$	B
20. $H = 2C - 2$	B
21. $H = 2C - 6$	B
22. $H = 2C - 4$	B
23. Number of contiguous methylenes	N
24. Methyl beta to $C=C$; $C=C-C-CH_3$	B
25. Isopropyl presence	B
26. Number of rings	N
27. Size of monocyclic	N
28. Smallest cycle	N
29. Fused rings	B

[a] N, Numeric; B, binary.

166

TABLE 10 Substructure Descriptors

30 CH₂=CH—CH₂—

31 CH₂=CH—CH₂—CH₂—

32 CH₃—C=CH—
          |
         CH₃

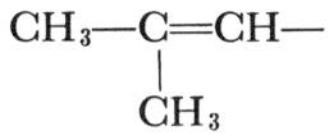

33 CH₂=C—CH₂—
         |
        CH₃

34 CH₂=CH—CH—
             |
            CH₃

35 CH₃—CH=C—
            |
           CH₃

36 CH₃—CH=CH—CH₂—

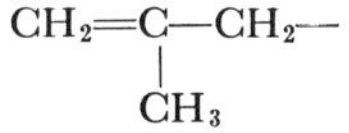

37   CH₃
          |
     CH₃—C—
          |
          CH₃

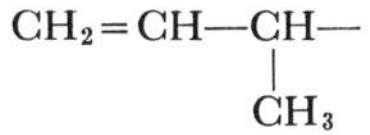

38   CH₃
          |
     CH₃—CH₂—CH—

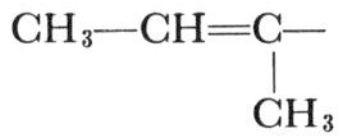

39 CH₃—CH₃—CH₃—CH₃—

40 CH₃
         |
     CH₃—CH—CH₂—

41 CH₃—CH—CH—
           | |
          CH₃ CH₃

42   CH₃
          |
     CH₃—C—CH₂—
          |
          CH₃

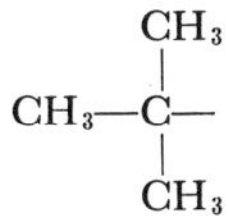

43 CH₃—CH₂—CH—
     CH₃—CH₂—

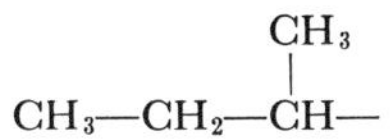

44   CH₃
          |
     CH₃—CH₂—C—
          |
          CH₃

45   CH₃
          |
     CH₃—C—
          |
          CH₃

46 CH₃ CH₃
     | |
    —CH—CH—

47 CH₃ CH₃
     | |
    —CH—C—
          |
          CH₃

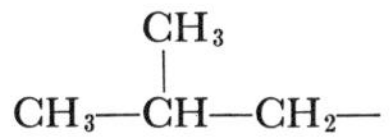

48 CH₃
     | CH₃
    —CH—CH₂—CH—

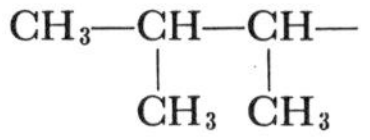

49   CH₃
          |
     CH₃ CH₂
      | |
     —CH—CH—

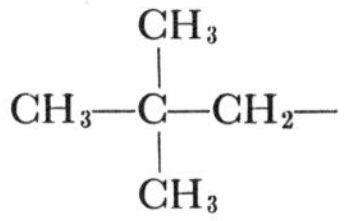

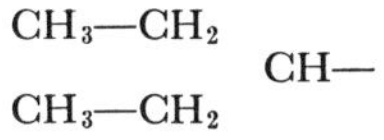

50 CH₃
     |
    —C—
     |
     CH₂
     |
     CH₃

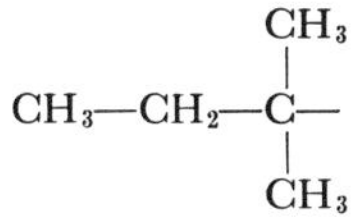

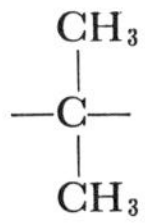

51   CH₃
          |
    —CH₂—C—CH₂—
          |
          CH₃

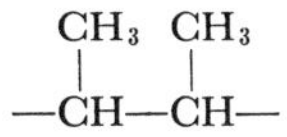

52 CH₃ CH₃
     | |
    —CH₂—CH—CH—

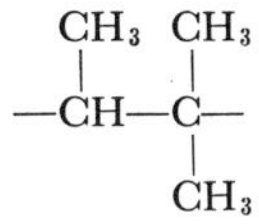

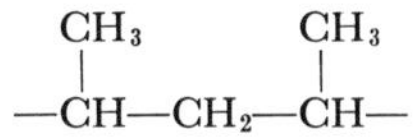

53   CH₃
                 |
    —CH—CH₂—CH₂—CH—
     |
     CH₃

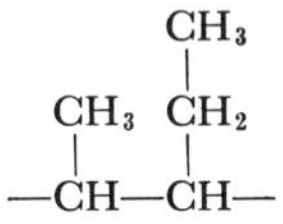

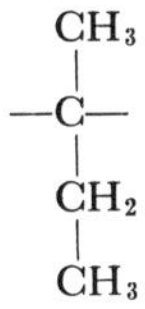

54   CH₂
          |
    —C—CH₂—CH—
     | |
     CH₃   CH₃

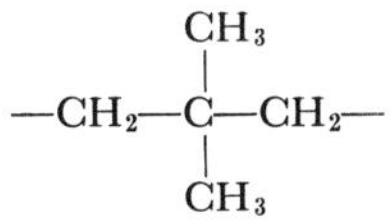

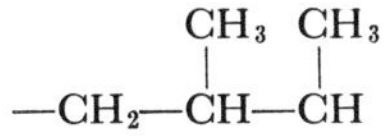

55 —CH—CH—CH—
      | | |
     CH₃ CH₃ CH₃

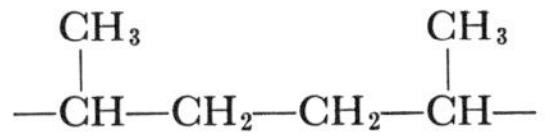

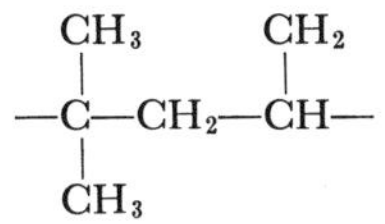

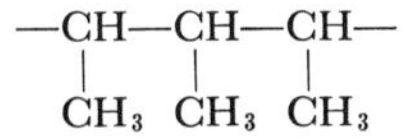

intensities range from 0.01 to 100.00% in each spectrum. The intensity of each peak is transformed into a logarithmic scale by the equation

$$I' = 10 \log (10000I)$$

in which I is the relative intensity of the peak as a percentage of the total ion current, and I' is the transformed intensity. The transformed intensities have values of 0 or are in the range of 10 to 60. In the logarithmic scale 30 corresponds to 0.1% of the total ion current, 37 corresponds to 0.5%, and 40 corresponds to 1.0%.

A single training session for an individual TLU proceeds as follows. The overall data set of 377 structures is coded in 55-dimensional vectors randomly divided into two subsets: a training set of 200 and a prediction set of 177. Then each subset has the structures removed for which the molecular weight is less than the m/e position being trained for. A set of correct responses is developed. For example, the TLU may be trained to determine if compounds have a mass spectral peak at $m/e = 57$ of magnitude greater than some cutoff intensity, say 35. Then the TLU will be trained using the 200 training set members so that it will give a positive result for those molecules having such a peak and a negative result for those lacking such a peak. After training is complete, the TLU is allowed to classify the (unknown) members of the prediction set. The percentage correctly predicted is reported as the predictive ability. In order to develop the ability to make decisions regarding the size of mass spectral peaks, several TLUs must be trained with different intensity cutoff values.

Table 11 shows the results of training TLUs for several different intensity cutoffs for each of 25 mass positions. The left column lists the mass positions for which TLUs were trained. For each mass position, three TLUs were trained with intensity cutoffs that divided the training set into 1:3, 1:1, and 3:1 populations. For each TLU three data are reported: (1) the intensity cutoff used, (2) the identity of the training set used; and (3) the predictive ability. Three different, randomly chosen training sets labeled A, B, and C were used. During training with any one of these training sets, if convergence was not obtained in 2500 feedbacks (an economic limitation), the five members of the training set that were most often missed were discarded and training was reattempted. The cycle could be repeated a second time if necessary. Thus the designation B-10 for the first TLU for mass position 29 means that the discarding cycle eliminated 10 structures from training set B, whereupon convergence occurred.

The predictive abilities obtained for all 73 TLUs ranged between 70 and 97%, with an average of 87.2%. The average predictive ability for the 1:3, 1:1, and 3:1 TLUs is given at the bottom of Table 11. Random guessing yields 50% predictive abilities for all the TLUs. Prediction based on the

TABLE 11 *Results of Training TLUs*

m/e	1:3			1:1			3:1		
	Cutoff	Training Set	Predictive Ability	Cutoff	Training Set	Predictive Ability	Cutoff	Training Set	Predictive Ability
29	41	B-10	94.9	45	B-10	83.6	47	A-10	91.5
41	47	C-5	93.8	49	A-10	69.5	50	B-10	84.8
42	38	C-10	89.3	41	A-10	76.0	44	A-10	81.1
43	35	B-10	93.2	43	A-10	87.4	47	C	96.6
54	31	C	95.4	36	B-10	93.1	40	C-10	84.0
55	40	B-5	91.4	46	A-5	90.3	50	A-5	82.3
56	33	B-5	92.5	42	C-10	86.1	48	A-10	80.8
57	29	B-5	86.8	37	C-5	89.0	46	A	95.4
66	25	A	92.9	31	A-5	82.8	34	B-5	93.5
67	31	C-5	90.5	37	B-10	89.4	42	B	97.1

68	25	B-10	89.4	33	B-10	94.7	39	B-10	88.2
69	31	C-5	90.5	40	A-5	88.8	46	A	84.0
70	28	C-5	95.0	39	B-10	82.9	45	B-10	81.1
71	19	B-10	84.2	31	B-5	82.9	38	B-5	89.6
80	19	B-10	90.6	25	C-10	86.4	30	B	91.3
81	22	A-5	90.1	29	A	92.1	38	B-5	96.3
82	18	C-10	85.0	27	B-5	91.3	36	C-10	83.7
83	24	B-5	85.6	32	A-10	77.3	41	B-10	87.5
84	21	C-10	86.4	30	C-10	81.6	41	C-10	78.9
85	21	B-5	80.4	27	C-10	72.8	35	A-10	83.9
95	1	B-5	85.1	19	A-5	93.0	31	B-5	87.2
97	19	B-10	79.4	26	B-10	75.9	34	A-10	85.9
98	21	B-5	82.8	28	A-5	78.9	41	A-10	83.7
105				1	B-5	90.8	31	B	92.7
106				1	B-5	88.9	24	C-5	93.2
Average			88.9			85.0			87.8

TABLE 12   *Feature Selection Results*

			55 Descriptors		Reduced Patterns			
m/e	Cutoff	Training Set	Feedbacks	Average Percent Prediction	Descriptors Retained	Fragments/ Substructures	Feedback	Average Percent Prediction
29	41	B-10	193/268	94.4	26	17/9	131/285	94.4
	45	B-10	1	71.5	42	23/19	1	78.8
	47	A-10	1946/1981	90.4	38	10/18	1888/2285	91.0
43	35	B-10	95/98	92.0	25	20/5	56/60	92.1
	43	A-10	1101/1048	87.1	39	21/18	1290/787	86.3
	47	C	2297/2276	95.8	38	22/16	2492/2501	95.8
67	31	C-5	2197/2504	91.1	42	23/19	2500/2505	91.6
	37	B-10	2005/1783	90.9	46	26/20	1857/1823	91.8
	42	C	2467/1464	92.0	27	22/5	867/384	93.2
68	25	B-10	346/570	90.8	27	18/9	349/323	92.0
	33	B-10	787/731	94.0	33	20.13	689/590	95.0
	39	B-10	309/229	90.5	23	18/5	161/169	91.4

known probabilities of the two classes $1:3$ and $3:1$ yields 50% for the $1:1$ class. The actual predictive abilities are far higher than this. Evidently, the TLUs have been able to distill some generalizations directly from the molecular structures about what makes structures give rise to mass spectral peaks of certain relative intensities.

Table 12 shows the results of applying a feature selection routine to the data sets used in Table 11. The feature selection routine attempts to discover which molecular structure descriptors are important in answering a specific chemical question. The procedure is as follows. For the particular mass spectral peak and intensity cutoff being investigated, two weight vectors are independently trained starting from different initializations. The feature selection process then discards those descriptors that contribute little or nothing to the solution of the problem. This is done by comparing the signs of the components of the two weight vectors. Only those descriptors for which both weight vector components agree in sign are retained. Then the two weight vectors of reduced dimensionality are retained. Feature selection continues until no unimportant descriptors remain.

Table 12 shows the feature selection results obtained for four mass positions. An average of 34 descriptors is retained. The number of descriptors retained for each TLU is shown in column 6, and the number of fragment and substructure descriptors remaining is shown in column 7. Comparison of columns 4 and 8 shows that training generally occurred either as fast or faster with a reduced number of descriptors compared to the entire set of 55 descriptors. The predictive abilities listed in the last column are almost always higher—the average predictive ability for all 12 questions is 1.1% higher with only six-tenths as many descriptors.

These results demonstrate that the information relevant to the chemical questions being asked resides in a fraction of the raw pattern vectors' components. By discarding the irrelevant descriptors, performance of the TLUs is enhanced.

REFERENCES

1. Joseph Schechter and P. C. Jurs, *Appl. Spectrosc.* **27**, 30 (1973).
2. Joseph Schechter and P. C. Jurs, *Appl. Spectrosc.*, **27**, 225 (1973).
3. P. C. Jurs, in *Computer Representation and Manipulation of Chemical Information*, W. T. Wipke et al., Eds., Wiley-Interscience, New York, 1974, p. 265.
4. H. M. Rosenstock and M. Kraus, in *Mass Spectrometry of Organic Ions*, F. W. McLafferty, Ed., Academic, New York, 1963.
5. M. L. Vestal, in *Fundamental Processes in Radiation Chemistry*, P. Ausloos, Ed., Interscience, New York, 1968.
6. H. M. Rosenstock, in *Advances in Mass Spectrometry*, Vol. 4, E. Kendrick, Ed., Institute of Petroleum, London, 1968.

7. B. Cuchanan, G. Sutherland, and E. A. Feigenbaum, in *Machine Intelligence* **4**, B. Meltzer and D. Michie, Eds., American Elsevier, New York, 1969.
8. R. W. Liddell, III, and P. C. Jurs, *Appl. Spectrosc.*, **27**, 371 (1973).
9. F. W. McLafferty, *Interpretation of Mass Spectra: An Introduction*, W. A. Benjamin, New York, 1966.
10. F. W. McLafferty, *Mass Spectral Correlations*, American Chemical Society, Washington, D. C., 1963.
11. S. K. Abdali, *Pattern Recognition*, **3**, 3 (1971).

Appendix. *Sample Learning Machine Program*

The example program consists of a short main program which inputs a set of data, initializes a number of parameters, calls the training subroutine, and calls the prediction subroutine twice.

The array DATA contains the raw data or patterns. It is dimensioned for up to 100 patterns of up to five components per pattern. The data are input from a set of cards and are assumed to be normalized already. W contains the weight vector being developed. LIST contains the category of each pattern, $+1$ or -1; this information is also input from the data cards. A total of NUM components per pattern is used. NTRSET is the number of patterns in the training set. IDTR contains the number of patterns in the prediction set, and IDPR contains a list of which members of the data set comprise the prediction set. The prediction set contains the entries NTRSET $+$ 1, NTRSET $+$ 2, . . ., NTOT in the example problem. NPASS contains the number of feedbacks that will be allowed before training is arbitrarily terminated because of lack of convergence. TSHD is the size of the dead zone or threshold with which the binary pattern classifier is to be trained. NCONV reports to the calling program whether convergence was attained (NCONV $=$ 0) or not NCONV $=$ 1). WINIT is the value to which each component of the weight vector is to be set before training begins.

Subroutine TRAIN uses a subsetting procedure. On the first pass it uses the entire training set, applying feedback whenever necessary. At the same time it keeps a list of which members of the training set were misclassified in NSS. On the second pass only those patterns missed in the first pass are classified, and a third subset is constructed. The process repeats until no patterns are misclassified. Then the entire training set is classified and the entire process begins again. The training program prints out the number of patterns misclassified on subsequent passes through the subsetting procedure.

An example execution is sown for an artificially generated set of data known to be linearly separable.

```
C          BASIC LEARNING MACHINE PROGRAM
C
       DIMENSION DATA (5,100),W(6),LIST(100),IDTR(100),IDPR(100)
       NTRSET=80
       NPRSET=20
       WINIT=0.1
       TSHD=0.75
       NTOT=NTRSET+NPRSET
       NPASS=1000
       NUM=5
C          INPUT DATA SET
       DO 10 I=1,NTOT
    10 READ (5,9) LIST(I),(DATA(J,I),J=1,NUM)
C          SET UP TRAINING SET
       DO 20 I=1,NTRSET
    20 IDTR(I)=I
C          SET UP PREDICTION SET
       DO 30 I=1,NPRSET
    30 IDPR(I)=I
C          INITIALIZE WEIGHT VECTOR
       DO 40 J=1,NUM
    40 W(J)=WINIT
       W(NUM+1)=WINIT
       CALL TRAIN (DATA,W,LIST,NTRSET,NUM,NPASS,TSHD,NCONV,IDTR)
C          CALL PREDICTION ROUTINE WITH DEADZONE OF 0.75
       CALL PRED (DATA,LIST,W,NUM,TSHD,NPRSET,IDPR)
       TSHD=0.0
C          CALL PREDICTION ROUTINE WITH DEADZONE OF 0.0
       CALL PRED (DATA,LIST,W,NUM,TSHD,NPRSET,IDPR)
     9 FORMAT (I5,5F10.3)
       STOP
       END
```

```
      SUBROUTINE TRAIN (DATA,W,LIST,NTRSET,NUM,NPASS,TSHD,NCONV,IDTR)
      DIMENSION DATA (5,100),W(6),NSS(100),KPNT(20),LIST(100),IDTR(100)
      NCONV=0
      WRITE (6,169) NTRSET,NUM,TSHD
      NUMM=NUM+1
      NF=0
      KNK=0
      KNV=0
C     START OF MAIN LOOP OF TRAINING PROGRAM (RETURN FROM ST 206)
   51 KKK=0
      IF (KNV) 54,54,53
   53 NDSS=KNV
      GO TO 65
   54 NDSS=NTRSET
      DO 60 I=1,NTRSET
   60 NSS(I)=IDTR(I)
C     THE 200 LOOP CLASSIFIES THE NDSS MEMBERS OF THE CURRENT SUBSET
   65 DO 200 IR=1,NDSS
      I=NSS(IR)
C     THE 70 LOOP CALCULATES THE DOT PRODUCT
      S=W(NUMM)
      DO 70 J=1,NUM
   70 S=S+DATA(J,I)*W(J)
C     THE NEXT THREE IF STATEMENTS TEST FOR THE CORRECT ANSWER
      IF (LIST(I)) 95,95,96
   95 IF (S+TSHD) 200,200,116
   96 IF (S-TSHD) 115,115,200
C     STATEMENT 115 OR 116 CALCULATES C, THE CORRECTION INCREMENT
  115 C=2.0*(TSHD-S)
      GO TO 117
  116 C=2.0*(-TSHD-S)
  117 XX=1.0
      DO 120 J=1,NUM
  120 XX=XX+DATA(J,I)**2
      C=C/XX
C     THE 130 LOOP PERFORMS THE FEEDBACK
      DO 130 J=1,NUM
  130 W(J)=W(J)+C*DATA(J,I)
      W(NUMM)=W(NUMM)+C
      KKK=KKK+1
      NSS(KKK)=I
      NF=NF+1
  200 CONTINUE
      KNV=KKK
      KNK=KNK+1
      KPNT(KNK)=KNV
      IF (KNK-20) 205,203,203
  203 WRITE (6,159) KPNT
      KNK=0
C     STATEMENT 205 TESTS FOR EXCESS NUMBER OF FEEDBACKS
  205 IF (NF-NPASS) 206,211,211
C     ST 206 TESTS FOR WHETHER CURRENT SUBSET IS ENTIRE TRAINING SET
  206 IF (NDSS-NTRSET) 51,207,51
C     ST 207 TESTS FOR WHETHER ZERO ERRORS WERE COMMITTED
  207 IF (KNV) 51,212,51
  211 NCONV=1
C     SUMMARY OUTPUT OF TRAINING ROUTINE
  212 IF (KNK.GT.0) WRITE (6,159) (KPNT(K),K=1,KNK)
```

```
      WRITE (6,179) (W(J),J=1,NUMM)
      WRITE (6,149) NF
  149 FORMAT (1H0,10X,9HFEEDBACKS,I6)
  159 FORMAT (1H ,20I4)
  169 FORMAT (1H0,10X,8HTRAINING,2I10,F10.2,/)
  179 FORMAT ('0',10X,'WEIGHT VECTOR',//,(' ',F17.3))
      RETURN
      END




      SUBROUTINE PRED (DATA,LIST,W,NUM,TSHD,NPRSET,IDPR)
      DIMENSION DATA(5,100),W(6),LIST(100),IDPP(100)
      LW1=0
      LW2=0
      KW=0
      NPA=0
      NNA=0
      DO 120 II=1,NPRSET
      I=IDPR(II)
      S=W(NUM+1)
      DO 50 J=1,NUM
   50 S=S+DATA(J,I)*W(J)
      IF (ABS(S)-TSHD) 101,102,102
  101 KW=KW+1
      GO TO 120
  102 IF (LIST(I)) 103,103,105
  103 NNA=NNA+1
      IF (-S-TSHD) 104,104,120
  104 LW1=LW1+1
      GO TO 120
  105 NPA=NPA+1
      IF (S-TSHD) 106,106,120
  106 LW2=LW2+1
  120 CONTINUE
      WRITE (6,109) TSHD
      LWT=LW1+LW2
      JW=NPA+NNA
      PW =100.0-FLOAT(LWT)/FLOAT(JW )*100.0
      PW1=100.0-FLOAT(LW1)/FLOAT(NNA)*100.0
      PW2=100.0-FLOAT(LW2)/FLOAT(NPA)*100.0
      WRITE (6,9) JW,KW,LWT
      WRITE (6,119) LWT,JW,PW,LW1,NNA,PW1,LW2,NPA,PW2
    9 FORMAT ('0',I10,'  NUMBER PREDICTED',/,' ',I10,'   NUMBER NOT PREDI
     1CTED',/,' ',I10,'   NUMBER PREDICTED INCORRECTLY')
  109 FORMAT (1H0,///,' PREDICTION WITH DEAD ZONE',F10.4)
  119 FORMAT (1H0,3(I10,1H/,I3,1X ,F6.2,5X))
      RETURN
      END
```

1	9.446	4.652	5.089	6.715	7.659
-1	1.047	7.768	1.232	9.566	9.657
-1	1.008	5.689	2.309	9.151	5.504
1	2.275	7.488	5.846	3.161	6.999
1	8.429	8.758	6.035	4.489	3.155
1	4.399	6.073	7.395	1.627	2.080
1	6.291	8.199	1.430	1.651	5.536
1	6.674	1.946	5.121	6.372	1.194
1	1.749	4.944	8.628	2.311	5.513
-1	8.184	2.320	1.038	9.559	3.555
-1	5.382	1.254	7.743	9.787	3.955
1	3.048	6.231	5.220	9.610	3.046
-1	2.261	4.320	4.996	7.056	9.340
-1	3.579	2.974	1.185	7.190	6.958
-1	4.085	1.623	3.737	7.432	5.965
-1	2.146	4.247	1.190	4.719	9.541
1	2.908	4.805	3.870	5.941	5.363
1	6.922	4.930	1.826	2.480	9.829
1	7.476	3.589	9.231	2.259	5.733
-1	1.177	1.362	1.915	8.731	6.713
-1	1.157	4.540	4.673	7.268	6.750
-1	2.749	1.998	1.341	5.190	8.152
-1	8.958	2.086	2.452	9.245	5.935
-1	2.093	1.533	9.708	8.097	7.790
-1	3.330	3.770	1.455	9.212	8.542
1	5.756	9.958	7.919	7.227	1.535
1	4.175	2.696	3.936	3.557	5.674
1	2.871	9.425	4.766	7.312	9.847
1	8.577	5.058	1.047	7.018	3.870
-1	2.654	3.293	7.049	9.879	3.001
-1	3.726	2.921	4.516	9.968	4.082
-1	3.092	1.436	9.442	9.286	4.120
1	4.348	6.583	5.728	2.982	2.992
1	6.689	8.739	4.492	8.297	3.041
-1	4.378	1.066	2.362	7.397	5.232
-1	8.009	2.085	1.943	9.197	4.167
1	9.455	7.800	2.158	2.379	9.298
1	9.266	9.077	2.034	7.627	6.522
1	5.503	4.550	6.226	4.038	8.867
1	7.345	1.293	3.187	3.292	2.449
-1	1.101	3.529	1.657	5.419	4.061
1	2.352	8.532	1.901	8.235	8.630
1	3.109	7.563	3.125	2.462	8.009
1	4.075	9.100	4.393	7.727	5.511
1	2.555	4.735	5.885	5.698	7.299
1	6.464	3.300	6.113	6.721	3.521
1	1.567	9.808	5.330	9.532	3.487
-1	7.158	1.457	1.034	9.176	9.725
-1	1.317	1.700	2.500	9.372	5.807
-1	2.056	4.475	1.524	6.913	7.857
-1	3.025	1.310	6.264	6.598	5.429
-1	1.737	8.103	2.569	9.755	9.521
-1	3.501	2.109	3.246	9.058	3.312
1	8.783	5.136	6.777	1.782	9.780
1	6.609	4.111	4.837	9.836	3.968
-1	2.221	4.699	3.316	7.243	6.017
1	6.879	5.609	3.993	3.646	3.119
1	9.439	7.733	6.781	4.484	9.830
1	1.654	7.676	7.608	3.614	2.206
-1	2.279	1.586	9.509	9.885	2.946

-1	2.623	1.391	4.690	8.046	8.887
1	3.734	5.677	5.274	7.038	3.660
1	2.135	9.473	8.350	9.809	5.401
1	6.344	3.960	4.158	3.321	5.287
-1	5.128	2.969	1.016	9.746	8.041
-1	2.153	2.851	8.124	9.884	5.044
-1	4.951	3.286	4.621	9.574	3.964
-1	3.533	1.038	9.379	8.235	5.738
-1	1.274	2.449	3.598	7.628	2.649
-1	2.007	3.298	2.301	9.787	1.103
-1	5.083	1.162	3.390	8.351	9.660
1	2.006	8.591	9.868	7.118	9.357
1	1.592	9.357	9.688	3.748	8.289
1	3.594	9.431	9.278	8.999	1.018
-1	2.302	2.236	1.904	8.819	3.910
1	9.445	4.268	3.223	8.712	7.521
1	2.260	4.152	6.866	1.817	9.250
-1	2.200	1.134	2.263	3.299	7.722
-1	1.737	1.964	2.830	5.420	8.458
-1	1.202	4.833	2.748	9.669	6.601
-1	3.478	3.282	1.898	5.178	8.770
-1	6.294	1.508	4.888	9.868	1.885
1	7.725	6.020	4.718	8.456	1.520
-1	1.394	1.421	2.657	6.356	6.314
1	2.125	2.623	2.794	1.176	2.070
-1	2.297	1.351	3.578	8.646	4.493
-1	4.130	2.838	8.205	9.652	6.082
-1	5.227	3.246	1.928	7.250	4.956
1	7.270	8.248	1.466	5.155	6.818
1	2.828	1.552	7.244	5.567	4.739
-1	4.458	1.216	6.749	8.130	5.664
1	5.712	8.347	3.220	9.281	1.201
1	5.004	9.805	4.983	9.737	8.059
-1	4.399	2.167	2.067	5.227	9.735
-1	4.763	1.766	5.146	9.342	9.912
1	9.608	3.267	7.283	9.118	6.363
1	2.467	5.562	4.090	3.670	2.879
1	6.401	1.149	8.228	1.455	4.615
-1	1.562	3.538	3.987	6.758	4.492
1	4.134	2.049	8.256	2.448	7.241

178

```
       TRAINING        80        5       0.75
 21  10   8   6   4   4   3   3   3   3   3   3   3   2   2   2   1   0  12   8
  7   6   4   4   3   3   3   3   3   3   2   0   6   5   4   4   4   4   3   3
  3   3   3   3   3   2   0   2   2   2   1   0   6   2   2   1   0   4   3   1
  0   3   2   2   1   0   4   3   3   3   3   3   2   2   2   2   1   0   0

       WEIGHT VECTOR

          0.605
          0.960
          0.299
         -0.714
         -0.410
          0.022

       FEEDBACKS    254


PREDICTION WITH DEAD ZONE     0.7500

       20   NUMBER PREDICTED
        0   NUMBER NOT PREDICTED
        0   NUMBER PREDICTED INCORRECTLY

       0/ 20 100.00              0/  9 100.00              0/ 11 100.00


PREDICTION WITH DEAD ZONE     0.0

       20   NUMBER PREDICTED
        0   NUMBER NOT PREDICTED
        0   NUMBER PREDICTED INCORRECTLY

       0/ 20 100.00              0/  9 100.00              0/ 11 100.00
```

179

Bibliography

1. H. C. Andrews, *Computer Techniques in Image Processing*, Academic, New York, 1970.
2. H. C. Andrews, *Introduction to Mathematical Techniques in Pattern Recognition*, Wiley-Interscience, New York, 1972.
3. D. A. Bell, *Intelligent Machines. An Introduction to Cybernetics*, Blaisdell, New York, 1962.
4. N. Bongard, *Pattern Recognition*, Spartan-Macmillan, New York, 1970.
5. G. C. Cheng, R. S. Ledley, D. K. Pollock, and Azriel Rosenfeld, Eds., *Pictorial Pattern Recognition*, Thompson, Washington, D. C., 1968.
6. N. L. Collins and Donald Michie, Eds., *Machine Intelligence 1*, American Elsevier, 1967.
7. Ella Dale and Donald Michie, Eds., *Machine Intelligence 2*, American Elsevier, 1968.
8. R. O. Duda and P. E. Hart, *Pattern Classification and Scene Analysis*, Wiley, New York, in press.
9. E. A. Feigenbaum and Julian Feldman, Eds., *Computers and Thought*, McGraw-Hill, New York, 1963.
10. L. F. Fogel, A. J. Owens, M. J. Wals, *Artificial Intelligence through Simulated Evolution*, Wiley, New York, 1966.
11. Keinosuke Fukunaga, *Introduction to Statistical Pattern Recognition*, Academic, New York, 1972.
12. K.-S. Fu, *Proceedings of the First International Joint Conference on Pattern Recognition*, IEEE, New York, 1973.
13. L. N. Kanal, *Pattern Recognition*, Thompson, Washington, D. C., 1968.
14. P. A. Kolers and Murray Eden, Eds., *Recognizing Patterns. Studies in Living and Automatic Systems*, MIT Press, Cambridge, Mass., 1968.
15. W. S. Meisel, *Computer-Oriented Approaches to Pattern Recognition*, Academic, New York, 1972.
16. Bernard Meltzer and Donald Michie, Eds., *Machine Intelligence 4*, American Elsevier, New York, 1969.
17. Bernard Meltzer and Donald Michie, Eds., *Machine Intelligence 5*, American Elsevier, New York, 1970.

18. J. M. Mendel and K. S. Fu., Eds., *Adaptive Learning and Pattern Recognition Systems*, Academic, New York, 1970.

19. Donald Michie, Ed., *Machine Intelligence 3*, American Elsevier, New York, 1968.

20. Marvin Minsky and Seymour Papert, *Perceptions*, MIT Press, Cambridge, Mass., 1969.

21. N. J. Nilsson, *Learning Machines. Foundations of Trainable Pattern-Classifying Systems*, McGraw-Hill, New York, 1965.

22. N. J. Nilsson, *Problem-Solving Methods in Artificial Intelligence*, McGraw-Hill, New York, 1971.

23. E. A. Patrick, *Fundamentals of Pattern Recognition*, Prentice-Hall, Englewood Cliffs, N. J., 1972.

24. G. S. Sebestyen, *Decision-Making Processes in Pattern Recognition*, Macmillan, New York, 1962.

25. J. R. Slagle, *Artificial Intelligence. The Heuristic Programming Approach*, McGraw-Hill, New York, 1971.

26. J. T. Tou and R. H. Wilcox, *Computer and Information Sciences*, Spartan, Washington, D. C. 1964.

27. J. T. Tou, Ed., *Computer and Information Sciences II*, Academic, New York, 1967.

28. Leonard Uhr, *Pattern Recognition*, Wiley, New York, 1966.

Index